Philosophy
of Chance

MICHAEL HELLER

Philosophy of Chance

A cosmic fugue with a prelude and a coda

Translated by
RAFAŁ ŚMIETANA

Copernicus Center PRESS

Originally published in Polish as *Filozofia przypadku*
by Copernicus Center Press, Kraków 2011

Translated by
Rafał Śmietana

Editing & proofreading by
Aeddan Shaw

Cover design:
Mariusz Banachowicz

Typesetting:
Byfflo Books

Publication supported by The John Templeton Foundation Grant
"The Limits of Scientific Explanation"

ISBN 978-83-7886-383-0

2nd edition. Kraków 2018

Publisher: Copernicus Center Press Sp. z o.o.
pl. Szczepański 8, 31-011 Kraków
tel/fax (+48) 12 448 14 12
e-mail: marketing@ccpress.pl
www.en.ccpress.pl

Printing and binding by OSDW Azymut

Table of Contents

Foreword

If I were to dedicate this book to someone, it would be to Richard Dawkins and William Dembski. They are worlds apart, yet they have even more in common. They stand divided by their views on the theory of evolution. Dawkins considers it to be a "blind watchmaker" who explains everything, while in Dembski's view it is full of "irreducibly complex situations" that testify to Intelligent Design. They are linked by their passion in the promotion of their respective views, by a willingness to stake everything on a single card. In the book that I now submit to the Reader, I put forward my own position, equally distant from the views of one and the other. However, I do not wish to join the club of zealots in preaching my own ideas, therefore I attempt to consider the issue from a broader perspective and in a broader context, while treating the discussion with both above-mentioned authors only marginally, by way of exemplifying their relevance to the problems at hand.

My broader context is constituted by the "philosophy of chance" not only (or perhaps, not principally) in the theory of biological evolution, but also in the structure and evolution of the entire Universe, and my broader perspective is the historical perspective. Even though I do not intend to write a history of the concept of chance (if I did, it would have to be a work of considerably greater length), I wish, however, to focus on at least some episodes in the history of science and philosophy, namely those that have played a more important role in the formation of the concept and its transformations. History explains a number of problems, including those with which we struggle even today, often only because it never occurred to us that we should have examined their development through the ages.

Looking at the evolution of the concepts connected with probability and chance, at the very outset we can see why we tend to associate the concept of chance (at this stage understood only intuitively) with the collapse of rationality. This association has made the process of domesticating random events considerably more difficult. After all, how one can understand something that is not rational? Help – yet over a long timeframe – arrived from two sources: from theological considerations – because chance may not break free from the embrace of Providence, and from the practice of gambling – because if one wants to win, one must somehow learn to control the randomness of chance. Thanks to the human passion for gambling, the exchange of letters between Pascal and Fermat constituted the foundations for what became later known as probability calculus, but soon this current was joined by theological strands at a key juncture: the notion of probability itself drifted in from theological disputes of Spanish casuists to the nascent probability calculus. The pace of events has since accelerated. Although in his seminal work Jacob Bernoulli still writes a lot about games of chance and hazard, they are more and more clearly replaced by purely mathematical theorems. Moreover, the practical applications of probability calculus testify to its maturity. For instance, mortality statistics have allowed the problem of epidemics in large cities to be overcome (since it was mostly there that reasonably accurate registers of the dead were maintained), and the application of statistical methods to the system of annuities began to bring profits to bankers. Thus ended the first stage of domesticating chance and thus ends the first part of this book.

Then there is a gap. If I wanted to write a history of probability calculus, I would have to add here several chapters detailing what happened between Laplace (inclusive) and the beginning of the 20th century. No doubt, this would be interesting and might even shed quite a lot of light on a number of detailed issues, but would distance me from the main objective of this book – the philosophy of chance. For this very reason, I decided to jump straight to the 20th century. The beginning of the previous century saw the emergence in mathematics of processes with a fundamental impact on the understanding of chance and related concepts. Probability calculus quickly transformed from a collection of statements and rules useful in various applications into a theory of modern mathematics. Thanks to Andrey Kolmogorov, who noticed the relationship between probability and the mathematical measure theory, and his formalisation of probability calculus, the concept of probability was at last liberated from intuitive associations and became an effective tool for conducting formal analyses. Thanks to this development, probability theory as a special case of measure theory, entered into fruitful interactions with other mathematical theories, rising to the role of one of the key branches of contemporary mathematics. Interpretative problems, so important for the 'philosophy of chance' did not disappear, but even here formalisation proved to be helpful in that it made it easier to perceive the boundary between the purely formal structure and its interpretative circumstances.

Progress in mathematics usually leads to generalisations. The fact that probability appeared in quantum mechanics in a new role, so to speak, not only intensified interpretative disputes, but also became a harbinger (or even a preliminary stage) of a generalisation of this concept. The process, once initiated, continued apace. Today, we have at our disposal a number of generalisations of probability theory (or, more accurately, probabilistic measure) of which the most spectacular include the so-called noncommutative probabilistic measures. Their particular beauty consists in that generalised concepts of probability possess a host of properties which are completely surprising from the vantage point of our mental habits.

The concept of chance is evidently connected with the notion of probability. An event is called random if we believe that its *a priori* probability of occurrence is small (at any rate, less than one). If we agree to adopt this intuition as the basis for a definition of chance, then we are immediately confronted by a question: What kind of probability are we talking about? In day-to-day situations, we are in no doubt that we mean

standard probability, more or less as defined by Kolmogorov. But in philosophical and worldview discussions it is impossible to avoid descend-
ing to the basic level (below the Planck threshold), because there – in our
opinion – the issue of the structure of the Universe 'is decided₁.' If this
level is indeed probabilistic, as all signs appear to indicate, in what sense
is it probabilistic? After all, there are many probabilistic measures. Both
the understanding of chance and its part in the structure of the Whole depend on our answer to this question.

In this way, almost imperceptibly, we have moved on to the third
part of the book where the problem of the 'philosophy of chance' is
brought to the fore. We thus return to the clash between Dawkins and
Dembski, which will help us outline our own solution. Let us formulate
the problem in a contrasting manner, as is done by the proponents of Intelligent Design: God or pure chance? This time, however, we approach
the question armed with a substantial dose of information concerning
probability calculus and equipped with an understanding of chance in
its context. We must, naturally, supplement our knowledge with a certain quantum of information from the history of theology, for it turns out
that the former masters, even though their knowledge in many domains
did not match ours, had a keen theological intuition (which we have, to a
large extent, squandered somewhere along the way). We will try to link
one with the other – our knowledge with a theological sense drawing on
tradition.

The existence of chance events in the structure of the Universe is
an irrefutable fact. They are not exceptional. They are interwoven into
the cosmic fabric. But their distribution in this structure is not accidental. They constitute an essential part of 'the matrix of the Universe.' They
appear as initial conditions for certain laws of physics and as fluctuations
that attack different dynamic processes occurring throughout the Universe. Without initial conditions, the laws of nature would not be able
to operate, moreover, without external fluctuations, non-linear dynamic processes would be in no position to generate authentic new developments. Biological evolution is such a dynamic process, with natural
selection, one of its basic mechanisms, consisting in an interaction be-

tween internal sensitivity of an evolving system with small fluctuations of the environment. Biological evolution constitutes one of the strands of cosmic evolution.

This extraordinary symphony of cosmos can be approached from different points of view. Like Dawkins, we may attempt explain everything by referring to pure chance. Like Dembski, we may try to discern the interventions of the Intelligent Designer in the exceptionally intricate details of the cosmic structure. Nevertheless, both of these attempts are unsuccessful and require fairly extensive intellectual acrobatics to maintain. Chance occurrences explain nothing, because they themselves demand an explanation. They are so subtly intertwined with the cosmic structure that without it they lose their import and cannot exist. And the reference to special places in the structure of the Whole as traces made by the Designer is, for this Designer, an insult exactly for the same reason for which the explanatory power of pure chance must be rejected. In the Cosmic Matrix there are no special places (regardless of whether we call them effects of chance or results of an extraordinary intervention); everything is part of the one Great Matrix. I would call it Intelligent Design, but this beautiful name has been compromised, therefore I prefer to use a phrase so often used by Einstein: "the Mind of God". And the purpose of science is nothing else but to decipher this "Mind of God".

***

I wish to express gratitude to my friend Jan Jaworowski, professor emeritus of mathematics at Indiana University, Bloomington, for his willingness to read a computer printout of this book. His precious comments permitted me to eliminate a number of inaccuracies, even errors. Those that remain notwithstanding, are my sole responsibility.

Every book consists of matter and form. Matter is the content, whilst the form is comprised of the words and sentences that are used to convey it. An ideal translation should preserve the content, while endowing it with a new form. And yet form shapes content, in a manner somewhat reminiscent of the philosophy of Aristotle, where it is the former that determines the heart of the matter. How is one, then, supposed to create something new while remaining faithful to the original? This challenge was faced by my translator, Rafał Śmietana, and his assistant Aeddan Shaw. I would like to express my heartfelt gratitude to them for meeting the challenge with such skill.

Tarnów, May 17, 2011

PART ONE

THE PRELUDE

A prelude can be an independent musical form, but – in a way, by its own nature – it tends to be an introduction to a larger piece. Often, it provides a prologue to a chamber sonata or a suite. Admittedly, the beginnings were modest, but history shows that a prelude will develop into something really magnificent – into something that, with respect to its craftsmanship and flourishes, I can only compare to a fugue. But let us not venture too far at this stage. We are at the very beginning. The journey will be long and demanding, though not devoid of truly artistic episodes.

In Antiquity and in the Middle Ages (Chapter 1), the notion of probability (as we understand it today) did not yet exist, so chance occurrences were discussed intuitively. People tried to curb the concept, wrapping it first in philosophical discourse, and then also in a theological one. The concept acquires its full meaning in contact with other notions. Chance was partnered by notions of causality and contingency. And they are here to stay: these affinities – regardless of whether they are based on similarity or contrast – will always be included in the set of images that until the present, we have associated with the concept of chance. But there also exists another face of chance. It came to light first in lists of statistical data. Random occurrences are organised in sequences whose subsequent terms can be anticipated and measurable advantages drawn from this fact.

Randomness and hope connected with forecasting meet in gambling, where the desire for profit becomes a powerful driving force. When Chevalier de Méré, the French nobleman and gambler, asked

Pascal for advice on how to solve a certain gambling puzzle, probability calculus found its mathematical development path.

Since then, events have happened more quickly (Chapter 2). In
our lives, we often make use of suppositions based on probability.
Should we do this using only our intuition and instinct? Logicians
from the Port Royal circle made an attempt at systematising these intuitions and expressing them in terms of their logical implications.
From there, only a single step separates us from the game with the
highest stakes, for example Pascal's famous wager, whether it makes
sense to risk choosing "God does not exist," since are two possibilities, either "God exists" or "God does not exist." Next, Huygens
takes over: he tries to summarise and develop all existing knowledge
concerning expectation (*expectatio*). He used this term in a technical
sense, later to be replaced by the term "probability."

The first genuinely mathematical study in the area of probability calculus, the posthumously published work by Jacob Bernoulli,
deserves more attention. Here for the first time appeared a theorem
complete with its proof (the first limit theorem), always true, irrespective of its context. Bernoulli's theorem deals with predicting the
results of long (theoretically, infinitely long) sequences of random
events but are random occurrences really determined? Bernoulli was
a zealous Calvinist and for him the question "How far does predestination reach?" was of vital importance.

Finally, in Bernoulli's work the term "probability" was first used
in its technical sense. Here as well mathematics is indebted to theology. Previously, the term had been used for a long time by theologians-
moralists who wondered to what extent was it ethical to be guided in
one's conduct by premises which are only probable, and how to measure the degree of their probability. Bernoulli knew these theological
disputes well and he transferred the issue of measuring the degree of
probability into his own mathematical reflections.

In mathematics, theory naturally leads to practical applications.
Bernoulli considered them, but he lacked the statistical data on which
to test his calculations. Such data were afforded to subsequent re-

searchers... by life itself. The lists of people who had died of the plague and banking annuities not only supplied input material for the theory, but also became a source of material gains when properly used.

Almost every abundantly fruitful domain of science sooner or later stimulates philosophical interest. The same was the case with probability calculus. When it was making its advances, the views later known as physico-theology, were quite common. According to its precepts, a number of constituent elements of the world's structure and operation (e.g. regular motion of planets, the structure of the eye as a complex optical instrument etc.) presupposed the existence of a rational Creator. However, can all these intricate details not be credited to chance operating in compliance with the rules of probability calculus? Pierre-Louis de Maupertuis dared to formulate such a supposition. He supplemented it with an important observation: chance in general produces undesirable, even monstrous results, but they are later invariably eliminated. Only those survive which, one way or another, prove to be advantageous. Was Maupertuis a forerunner of the idea of natural selection? Was Darwin familiar with his work? In the domain of physics, Maupertuis discovered the principle of least action, but endowed it with universal import. He thought that the Creator did not deign to design specific solutions, since He prefers to establish a general principle – the principle of least action – which imposes a master plan and within its framework, to coordinate all specific cases. De Maupertuis could have achieved much more, had he not been too hasty with sweeping generalisations and had he not become involved in intrigues meant to ensure to himself priority in the discovery of the least-action principle.

A brief recapitulation is now in order (Chapter 5). The long period of time leading up to the emergence of probability calculus can be interpreted as the time needed to domesticate or even control chance – from Aristotle, who maintained that chance constitutes a breach in rationality and therefore science remains powerless in its face, to Jacob Bernoulli, who formulated and mathematically proved his theorem on a long series of random events. Progress in science results in a gradual bringing into focus of methods, of our rationality, to the effect that what once used to be beyond its range, now becomes an object of rational study. But every expansion of rationality engenders new interpretative and philosophical issues. The journey is not yet finished. It is only a prelude.

Chapter 1

From Antiquity to Pascal and Fermat

1. Gambling in antiquity and afterwards

There is a stereotyped conviction that the history of probability calculus began with the famous exchange of letters between Pascal and Fermat of 1654. Ian Hacking[1] illustrates this with the following fact. In 1865, Isaac Todhunter published a book *A History of the Mathematical Theory of Probability from the Time of Pascal to that of Laplace*. This book became an authoritative survey of nearly all of the work in this domain for a long time. It is 618 pages long, but only 6 (six!) of them deal with what happened in probability calculus before Pascal. Hacking appears to consider this proportion to be justified.

> There was hardly any history to record before Pascal – he writes – while after Laplace probability was so well understood that a page-by-page account of published work on the subject became almost impossible.

Hacking[2] thinks that science develops thanks to two mechanisms: problems that it generates itself and problems that have been imposed on it from outside. Probability theory has posed problems for science itself to solve since quite recently, essentially since the early 20th century. In the times of Pascal and Fermat, the external factor that initiat-

---

[1] *The Emergence of Probability*, 2nd ed., Cambridge University Press, Cambridge 2006, p. 1.
[2] *Ibid.*, p. 4

ed scientific thinking about probability, was constituted by games of chance. It was only after Laplace that the motivation increased. The motivation evolved gradually deeper connections both with practical applications (insurance and annuities, needs of the economy) and scientific theories (statistical physics, measure theory and astronomy).

Yet games of chance have existed for as long as humanity itself. This general statement concerning antiquity is supported by extensive documentation. For example, it is known that the first dice were actually the ankle or heel bones of fast-running animals, such as horses or deer. Well-polished, and later also nicely adorned, they had the property of always falling on one of their four surfaces, once thrown. Such bones have been found in archaeological excavation sites in ancient Egypt; illustrations on tombs clearly indicate that they were used for gaming purposes. It is also known that Emperor Marcus Aurelius (also a philosopher) so often amused himself by throwing dice that he used to take along a special servant who played the role of his croupier on his journeys. Yet this early fondness for gambling did not generate a need for mathematical thinking about probability. Predecessors of Pascal and Fermat are not to be found in this area.

2. "What happens, does not happen always, perforce, or the most often"

Still, the Greeks were philosophers by nature. They were fascinated by speculations over the nature of the world and cognition rather than by games of chance. It is their philosophical inquiry that reflects the strands of thought which, over time, were to create a climate conducive to posing relevant questions concerning probability. Considerations regarding fate, chaos and chance occupy a prominent place in the Greek philosophical landscape; however, we will not follow the scattered remarks and opinions. Instead, let us invoke an author who explicitly and with reasonable systematicity poses the problem of chance.

In Book II of his *Physics*, Aristotle addresses the problem of chance in connection with his analysis of causes and causality since, in his opinion, it breaks away with causality. Some events "always come to pass in the same way," while "others, for the most part." However,

> It is clearly of neither of these that chance is said to be the cause, nor can the 'effect of chance' be identified with any of the things that come to pass by necessity and always, or for the most part.[3]

As was his habit, Aristotle begins his systematic analysis by enumerating the main positions on the issue of chance, of which, in his view, there are three.

> There are, thus, those who question the existence of chance.

> They say that nothing happens by chance, but that everything which we ascribe to chance or spontaneity has some definite cause.[4]

For example somebody, moved by a definite reason comes to the market where he meets the man whom he had wanted to meet. He calls this chance, although it is not a chance event, because that other man surely came to the market for a specific reason.

There are also those who assert that chance is the root cause of all, because

> They say that the vortex arose spontaneously, i.e. the motion that separated and arranged in its present order all that exists.[5]

---

[3] *Physics*, II, 196b. (All quotations below come from the edition by NuVision Publications, LLC, Sioux Falls 2007.)
[4] *Ibid.*, II, 196a.
[5] *Ibid.*

Here, obviously, Aristotle has the atomists in mind, but definitely does not share their position. He confesses that for him, "we should have expected exactly the opposite."[6]

There also exists a third group of thinkers who maintain that

chance is a cause, but that it is inscrutable to human intelligence, as being a divine thing and full of mystery.[7]

What is, then, the view of Aristotle himself? It follows from what, in his opinion, should be understood as chance. Below follows an appropriate source text (this time, from his *Metaphysics*):

We say that everything either is always and of necessity (necessity not in the sense of violence, but that which we appeal to in demonstrations), or is for the most part, or is neither for the most part, nor always and of necessity, but merely as it chances [...][8]

This leads to Aristotle's definition of chance: "The accidental, then, is what occurs, but not always nor of necessity, nor for the most part."[9] That is why "there is no science of such a thing; for all science is of that which is always or for the most part, but the accidental is in neither of these classes."[10] An essential corollary is to be found again in *Physics*:

Thus to say that chance is a thing contrary to rule is correct. For 'rule' applies to what is always true or true for the most part, whereas chance belongs to a third type of event. Hence, to conclude, since causes of this kind are indefinite, chance too is indefinite.[11]

---

[6] *Ibid.*, II, 196b.
[7] *Ibid.*
[8] *Metaphysics*, XI, 1064b. (All quotations below come from the translation by Rev. John Henry MacMahon, H.G.Bohn, London 1857).
[9] *Ibid.*, XI, 1065a.
[10] *Ibid.*
[11] *Physics*, II, 197a.

Thus, chance cannot be mathematised and finds itself entirely outside the domain of rationality. According to Aristotle, science by its own nature is causal, whereas chance constitutes a break of causality, and as such cannot be studied by science. Such a qualification of chance proved to be exceptionally long-lived. Its distinct traces can be found in an even today widespread belief that the biological theory of evolution cannot be reconciled with religious faith because the theory of evolution attributes a crucial role in the genealogy of life to chance, and chance constitutes a breach in the rationality of the plan of creation.

However, if chance breaks completely free from the authority of rationality, why can one still detect certain regularities in certain areas in which chance appears to reign supreme? Two such areas are medicine and the judiciary. In both domains, the Greeks developed considerable mastery in diagnosing and forecasting. Authors from the Hippocratic School created a kind of science based on – writes Crombie –connections and regularities of events that occur for the most part, although not in a constant or necessary manner, once one has observed their sufficient number. It offered objective descriptive knowledge which could be obtained by way of induction without the need to recognise the causes, by observation and recording those constant, random regularities.[12]

We often face a similar situation during judicial proceedings when, on the basis of circumstantial evidence, it is necessary to reconstruct a certain course of events. In speeches made to the court, as in each and every case of rhetoric, the point is, first of all, to persuade the listeners – judges in this instance. In his *Phaedrus*, Plato writes:

> in courts of law men literally care nothing about truth, but only about conviction: and this is based on probability, to which he who would be a skilful orator should therefore give his whole attention.[13]

---

[12] Cf. A. C. Crombie, *Styles of scientific thinking in the European tradition*, Duckworth, 1994.
[13] *Phaedrus*, Cosimo Classics, New York 2010, p. 85.

Let us leave the task of cataloguing and organising the Greek rules of reasoning based on symptoms and circumstantial evidence to historians of science.[14] I only wish to draw the Reader's attention to the fact that the rudiments of organised thinking about probability should be sought as far back as the antiquity. Even though the route from Greeks to Pascal and Fermat will still be long and arduous, the same applies to the route from Aristotle's *Physics* to Newton's .

Crombie argues that, for most ancient philosophers, the necessity to employ argumentation invoking probability followed from their uncertainty associated not so much with the very causality of nature, but with our knowledge about it[15]. It was only the Epicurean doctrine of random collisions of atoms, thanks to which the entire order of nature developed, that introduced "an internal indefiniteness to the Greek concepts of the nature of things." This was to remain the case for a long time to come. The belief in 'a necessary order of nature' would remain rooted in European thinking for a long time. It was not until the emergence of quantum mechanics that a huge breach was made in this train of thought.

3. From chance to contingency

The cosmos of the ancient Greeks was harmonious and orderly, because it was ruled by necessity. Although Aristotle warned that it was "not in the sense of violence, but that which we appeal to in demonstrations," in no way, however does this change the fact that the term *ananke* as used by Greek philosophers to designate the necessity inherent in nature, in its primary sense referred to coerced testimony, even by means of torture.[16] In this context – as we have seen

---

[14] Cf. e.g. A. C. Crombie, *op. cit.*
[15] *Ibid.*
[16] Cf. O. Pedersen, *The Two Books. Historical Notes on Some Interactions between Natural Science and Theology*, Vatican Observatory Foundation, Vatican City 2007, p. 9.

– chance was either something irrational or which stemmed from our ignorance (if we were to disregard the views of the atomists, but in Greece, they never constituted a serious alternative to more fashionable philosophies[17]).

Everything changed with the advent of Christianity. Christian theology emerged from the juxtaposition of thinking in accordance to the Bible with Greek philosophy.[18] Even if the Fathers of the Church introduced fundamental corrections into the Greek image of the world, they did this laden with a complete menu of Greek categories of thinking. As a result, the problem of the relation of chance to necessity remained. What had to occur first was the process of accent shifts followed by a gradual change in the meaning of individual terms and finally, a completely new understanding of the issue. Crombie, summarising result of our process of interest, says that the Providential theology of Creation, as revealed in Hebrew writings, assumed the existence of a transcendent Creator whose almighty freedom and unknowability made all events and their relationships definitely random from the human point of view. But within such randomness, conjecture concerning the possible relationships and their likely consequences found their basis in the certainty assuming a homogeneous causality in nature.[19]

On the one hand, the rationality of the Greek world was fully preserved, even strengthened by the reference to the rationality of the work of creation, yet, at the same time, this work ceased to represent the result of certain internal necessities. Instead, it became the result of the free intention of the Creator. Over time, Greek irratio-

---

[17] The views of atomists on the mutual relations of chance and causality are not entirely clear. In this context, Pedersen quotes a statement by Leucippus: "Nothing happens by chance (*maten*), but all occurs for a reason (*logos*) and by necessity (*ananke*). A more detailed commentary on the subject can be found in O. Pedersen, *op. cit.*, pp. 23–27.

[18] I write more extensively on the subject in *"Teologia a nauki przyrodnicze w okresie Ojców Kościoła,"* (Theology and the Natural Sciences in the Age of the Church Fathers, in Polish)*, in M. Heller, Z. Liana, J. Mączka, W. Skoczny, *Nauki przyrodnicze a teologia. Konflikt i współistnienie (Natural Sciences and Theology. Conflict and Coexistence)*, OBI – Biblos, Kraków – Tarnów 2001, pp. 29–51.

[19] Cf. A. C. Crombie, *op. cit.*

nal randomness was replaced by a new concept of contingency. The universe is contingent not only because God could have created it or not, but also because He could have created it to be one way or another. And here is where the storm of discussion broke out. Zealots, such as Petrus Damiani, maintained that God could do whatever He may wish, even change basic mathematical truths or the already completed past. Rationalists thought that the absence of any limitations leads straight into the embrace of contradiction. Anselm of Canterbury answered Petrus Damiani that God, who may do anything that He wants, might, for example, want to annihilate Himself. A way out of this dilemma was the distinction, already outlined by Origen, into the absolute (*potestas absoluta*) and orderly (*ordinata*) powers of God. Through His own absolute power, God may do anything that does not lead to contradiction. However, there are certain things that God could do by virtue of His own absolute power, but never will, because He is rational, just and good. These constraints determine the range of His orderly power. This distinction – which is only mentioned here – gradually matured and became clearer in the writings of Moses Maimonides, Alexander of Hales, Peter Lombard, in order to become fully crystallized in the works by Thomas Aquinas.

These theological disputes have their consequences in the study of the world. The constraints of God's omnipotence are imprinted – so to say – in the structure of Cosmos. The Greek Universe was extremely rationalistic; all knowledge about it could be, in principle, deduced from the first metaphysical rules. Chance in such a Universe was a foreign body, was merely tolerated and efforts were made to reduce it to the status of a factor of our ignorance exclusively. However, if God is free in his omnipotence (even if limited only to *potentia ordinata*), then all the properties of the world cannot be deduced from first principles, consequently, deduction needs to be supported or maybe even entirely replaced by watching attentively what the world is like, what God willed it to be like. As Amos Funkenstein[20] claims, it was one of the important factors which constituted a conducive climate to the emergence of empirical sciences. He also asserts that constraints on God's omnipotence transferred onto the world triggered the formation of the modern notion of natural law. The Universe seen through the prism of such a concept, so to say in principle contains room for events which cannot be explicitly anticipated. A contingent world is a world open to probability.

Let us note that the *ordinata* divine power also extends to include what God could have created, but did not. This theme was also present in scholastic discussions. When Leibniz spoke about the 'possible worlds' of which ours is the best, and which occupy in his philosophy such an important place, was an heir to that tradition. Naturally, over time, the methodological status of possible worlds changed. In scholastic discussions, they were theological possibilities, in Leibniz's work, they acquired a tinge of logical alternatives, in contemporary philosophy of science they constitute tools for counterfactual analysis of certain laws of nature, and in today's discussions around cosmology, they are becoming actually existing worlds. Concepts, once begotten, multiply manifold and mutate more quickly than biological species.

In the Middle Ages, the domains in which one was obliged to invoke probability involved not only speculations on the subject of God's omnipotence and contingency, but also the areas in which philosophical reflection overlapped with daily life. The necessity to act in specific situations connected with the moral postulate to be always guided by maximum certainty led to the emergence of different degrees of certainty. One Aristotelian principle was invoked here, namely that the kind of knowledge should be appropriate to its own object of study. In this respect, mathematical certainty differs from that of medicine and ethics. Medieval thinkers developed subtle classifications of reasonings and kinds of

---

[20] This subject is dealt with extensively in Chapter III of his book *Theology and Scientific Imagination form the Middle Ages to the Seventeenth Century*, Princeton University Press, Princeton 1986.

certainty that can be achieved on their basis. Crombie points to a statement by St Thomas Aquinas from his *Expositio super librum Boethii De Trinitate:*

> the more a science draws close to particulars (as do practical sciences like medicine, alchemy, and ethics), the less certain they can be because of the many factors to be taken into account in these sciences, the omission of any one of which will lead to error, and also because of their variability.[21]

Using today's language, this statement indicates the 'statistical dispersion' as one of the sources of obligation to invoke reasonings based on probability.

With reference to such classifications of reasoning, medieval thinkers developed a number of systematic rules that applied to the collection of data and evaluation of arguments in areas such as rules of conduct with heretics, infectious diseases or usury. In the latter case, analyses involved numerical comparisons. The issue was e.g. how the lender should calculate interest in proportion to the capital committed and the level of risk incurred. In the 14th century, such considerations entered the practice of commerce and the gradually developing insurance system for good. The transfer of such analyses onto the domain of games of chance was but a consequence of the process.[22]

4. Precursors

Gerolamo Cardano (1501–1576) was a passionate gambler. He read law and medicine at the universities of Pavia and Padua. As a student, he earned his living by casting horoscopes and teaching ge-

---

[21] A. C. Crombie, *op. cit.*

[22] For a more extensive treatment of the subject, cf. *ibid.* The universality of such 'probabilistic practices' in the Middle Ages is corroborated by the fact that on several pages of Crombie's book, footnotes documenting such practices occupy more space than his original text.

ometry and astronomy. Time and again, he also gambled. Since he had his ups and downs in life, he was not averse to supplementing his income in this way. He preferred games of chance, in which the odds were even for all of the players, to those which demanded a special strategy. Since his innate mathematical talent permitted him to estimate his chances better, he made quite a lot of money in the process. These experiences contributed to his short *Liber de ludo aleae* (*Book on games of chance*),[23] which was the first book in history devoted to the subject (actually, it is only a dozen or so pages long, but the author himself called it a book). In fact, it is a collection of notes made by Cardano over a number of years. They were collated and printed in his *Opera omnia* only in 1663, already after the publication of the famous correspondence between Pascal and Fermat, and this is the reason why Cardano's work, although historically earlier, did not play a greater role in the history of probability calculus.[24]

Cardano's short book is not homogeneous. Apart from what we would call today probabilistic considerations, it contains descriptions of period games of chance, almost personal notes and observations. Certain portions of his work can be interpreted as an attempt to defend gambling (Cardano extensively discussed not only the benefits but also the dangers of giving oneself over to gambling), although a portion of the book describes how to cheat at gambling. But the basic foundation adopted by Cardano in his probabilistic analyses is that the game is played fairly. Otherwise, mathematical rules of prediction do not apply. Cardano analyses in detail games that consists in throwing two and three dice. He realizes that the key role is played by the ratio of favourable results to all possi-

---

[23] Its English translation can be found in O. Ore, *Cardano, the Gambling Scholar*, Princeton University Press, Princeton 1953.

[24] In the history of mathematics, Gerolamo Cardano is better known for his contribution to algebra. His major work in this area was *Ars Magna* (1545). He was also the first to introduce the concept of complex numbers.

ble ones (but he does not call it probability); his analyses comprise enumerations of all of the possible results and calculations of the frequency of their occurrence.

Hacking[25] suggests that Cardano's analyses are based on his conviction that in sequences of random events, a certain role is played by the tendency for certain outcomes to occur more often than others. This would point to a certain analogy with the so-called propensity interpretation of probability advanced by Karl Popper, however – in Hacking's view – this analogy should not be pushed too far. Inasmuch as today's propensity interpretation has the task of somehow co-ordinating the randomness of events with the conception of causality imposed by quantum physics, the conception of causality applicable during Renaissance tended to have quite different associations. Gerolamo Cardano was first of all a physician (a very popular one in some periods of his life) and it is in his conceptions of medicine and nature where these causal representations should be sought.

An analysis of results of rolls of three dice (without a reference to Cardan) was undertaken by Galileo. The Grand Duke of Tuscany, his guardian and patron, asked him to solve the following problem: Why in throws of three fair dice does the number 10 appear more frequently than the number 9? The prince must have been a keen observer, having noticed this slight difference in the frequency of occurrence. A solution to this 'paradox' was described by Galileo in his short note *Considerazione sopra il Giuoco dei Dadi* (1718). Leonard Mlodinow, a present-day physicist and writer, explains the solution found by Galileo in the following way:

> If you throw a single die, the chances of any number in particular coming up are 1 in 6. But if you throw two dice, the chances of different totals are no longer equal. For example, there is a 1 in 36 chance of the dice totaling 2 but twice that chance of their totaling

---

[25] I. Hacking, *The Emergence...*, *op. cit.*, pp. 54–56.

3. The reason is that a total of 2 can be obtained in only 1 way, by tossing two1s, but a total of 3 can be obtained in 2 ways, by tossing a 1 and then a 2 or a 2 and then a 1.[26]

Galileo was not a gambler himself and he contemplated probability only as part of his 'commission,' but once the roulette wheel was put in motion, it keeps spinning... Fortunately, among gamblers there are those who also ask difficult questions.

5. The geometry of chance

Antoine Gombauld, Chevalier de Méré et Sieur de Baussay was a man of experience and a passionate gambler. Moreover, he must have had a flair for mathematics, although surely he was more interested in the strategy of the game and his chances of winning rather than in finding solutions to probabilistic puzzles. When he met Pascal during a social event, he posed a certain problem to him: A game of dice must be interrupted before the players can finish. How to divide the stake between the two players who cannot complete the game? After a moment's thought, Pascal answered that every player should receive an amount proportional to the probability that he would win the game if it were finished. Thus Pandora's box was opened: Pascal could not stop thinking about probability.

---

[26] Cf. L. Mlodinow, *The Drunkard's Walk: How Randomness Rules Our Lives*, Pantheon Books, New York 2008, p. 62. Let us, however, go back to the original question posed by the Grand Duke of Tuscany. It turns out that the probability of obtaining a 9 equals 25/216, while the probability of obtaining a 10 is 27/216 – a little more, but not much. The difference is due to the fact that 10 can be obtained as a sum of three numbers from 1 to 6 in 27 ways, whereas 9 can only be obtained in 25 ways. The difference occurs if the addends are numbers greater than 2. Then, $10 = 3 + 3 + 4 = 3 + 4 + 3 = 4 + 3 + 3$, whereas 9 can be represented only as a single sum of addends greater than 2: $9 = 3 + 3 + 3$. There are 216 possible threes of numbers from 1 to 6. There are 25 threes whose sum equals 9 and 27 threes whose sum equals 10. Hence, the probability of obtaining a 9 is 25/216, whereas the probability of obtaining a 10 is 27/216.

Pascal decided to consult Fermat on his considerations. He had never met him personally, but he had already exchanged letters with him (on inductive reasoning). The exchange of letters between Pascal and Fermat concerning probability lasted for four months in the summer and autumn of 1654. The correspondence comprises six letters in total. Unfortunately, the first letter is no longer extant. Pascal's biographer, William R. Shea, writes:

> Lady Luck may have been born in the slightly disreputable atmosphere of the gambling room, but she was given her titles of nobility in the fascinating exchange of letters between Pascal and Fermat [...][27]

We will not follow this correspondence as it is enough to focus on its main outcomes.

In the course of their exchange of letters, Pascal and Fermat invented three methods of 'taming chance.' The first was the result of their combined efforts – it is a combinatory method. Two players play a game with high stakes. The conditions for winning are known: the winner must obtain a certain number of points. The game is interrupted when one of the players has managed to obtain 2/3 of the score, whereas the other one – 1/3. The method consists in determining the maximum number of rolls necessary to indicate the winner. Then one should enumerate all possible series of results in these sequences, which will permit to indicate the winner.

Pascal, however, did not fancy the arduous task of enumerating all of the possibilities ('combinatorics') and for that reason proposed a method that very ingeniously avoided the need to do that. He called it the method of expectations. Still, we are facing the problem posed by Chevalier de Méré. A game of chance is inter-

---

[27] W.R. Shea, *Designing Experiments and Games of Chance*, Science History Publications, Watson Publishing International, Canton, Mass. 2003, p. 260. The correspondence between Pascal and Fermat and Pascal's contribution to probability calculus are extensively discussed in Chapters 11–13.

rupted and it is necessary to determine how the stakes are going to be split. Pascal's method consists in an attempt to somehow replace an uncertain future with a certain present. The contributions of the players to the prize pot cease to be their property. Pascal writes that the stake is the price paid by each player for the ability to expect a winning. As long as the game goes on, each player has a claim to a certain portion of money in the pool in proportion to previous success in the game. The stakes should be divided in accordance to this proportion if the game was interrupted. Chance, which would otherwise decide as to the result in the future, is replaced by a decision-making strategy concerning the present. Of course, this strategy – just as others developed in the correspondence – was accompanied by examples and computational technologies that permitted them to present their results in numerical form. To that end, Pascal made copious use of his famous arithmetic triangle (today called Pascal's triangle).

The third method, the direct probability method (thus called later; in the correspondence the word 'probability' is nowhere to be found), was proposed by Fermat. It consists of enumerating all the possibilities that may occur and assigning a fraction to every one of them. Today, such a fraction would be called the probability of a given event.

In what sense can the correspondence between Pascal and Fermat be considered the beginning of today's probability calculus? In the same sense as all mathematical theories can when they begin. First, there is a problem to be solved. Then – depending on the vagaries of history – after various attempts and more or less successful approximations, effective computational methods appear. Usually, a considerable time must elapse before they become integrated into an elegant mathematical structure. The theory matures when such a mathematical structure begins to interact creatively with other mathematical constructions.

In 1654, Pascal wrote to the Paris Academy of Sciences a letter in which he presented his results concerning the games of chance. They unite – he wrote – the randomness of chance with the exactitude of geometry and did not hesitate to call them "the geometry of chance," but the route to a fully-fledged mathematical theory was still to be a long one.

Chapter 2

Small and high stakes

1. Port-Royal Logic

A certain gentleman (*une personne de condition*) in conversation with a young man (Honoré d'Albert, Duke of Chevreuse) mentioned that in his youth, he had met someone who taught him almost the whole of logic within fifteen days. Another man, who had no great respect for the subject, was listening in on this conversation, and remarked that he would undertake this task himself within four or five days. The challenge was accepted. It should be noted, however, that even at the stage of the first draft, so many interesting things and new observations appeared that it would have been a great waste not to have compiled more detailed notes. These notes were then combined to form a book titled *La logique, ou l'art de penser*, commonly known as *Port-Royal Logic*. Its first edition was published in 1662 (two additional printings were soon to follow) and until 1683 it had had as many as five editions. Very soon, the book was translated into all major European languages.

In 1652, Blaise Pascal's younger sister, Jacqueline, took the veil and entered Port Royal Abbey, and soon afterwards – on November 23, 1654, to be precise – Pascal himself experienced a great mystical experience which altered his life.[28] All that strengthened his association with the Port Royal circles, which was an intellectually creative

---

[28] After Pascal's death, a note was found sewn into his clothes, in which he described his experiences of that night. Cf. W. R. Shea, *op. cit.*, pp. 194–196.

group with strong Jansenistic inclinations.[29] This is where *Logic or the art of thinking* stemmed from and, at first, the work was published anonymously, but soon names of its authors became widely known. They were Antoine Arnauld and Pierre Nicole, but the exact contribution of either remains unknown to this day. It is certain that Pascal was an influential force behind the book and he may have even written some parts of it himself.

Originally, the book was intended to serve as a kind of manual. The form was maintained, but the formula was considerably expanded: it was no longer an abridged version of logic to be learnt within several days, but rather an extensive guide to the art of thinking intended to accompany a young person throughout their life. The guide considerably exceeded the scope of what was then considered to be logic. It also included ample advice on how to move about in the world of thought, in the event of having to make decisions based on fragmentary premises in the absence of more complete arguments. No doubt, it was thanks to these features that *Port Royal Logic* gained a huge popularity.

2. Inspirations and content

The book is interspersed with prominent religious inspirations. The spirit of piety prevailing in the Port Royal circle is present almost everywhere, often emerging from implied meanings. The authors, imbued with religious faith, were well aware that it was not always possible to resort to irrefutable rational arguments. As a consequence, in such instances other kinds of justification should be applied. Even when in similar situations found in a 'secular context' the authors seek a proper strategy, the source of inspiration can be clearly discerned. The book consists of four parts: Part One is devoted to ideas,

---

[29] Jansenism was a theological and ascetic movement that emerged as a reaction to Reformation, but soon became more radical and was condemned by Pope Innocent X in 1655.

Part Two – to judgements, and Part Three – to models of reasoning. Part Four is entitled *Of Method* and, after preliminary comments concerning science in general, focuses on quite broadly conceived methods of analysis and synthesis.

The book is doubtless fascinating in many respects, but would be unlikely to deserve a mention in considerations devoted to the history of probability, were it not for several last chapters in Part Four. Interestingly, these chapters are missing from the manuscript preserved in Bibliothèque Nationale in Paris, neither are they listed in the table of contents. It appears that they were added 'at the last minute' before the book went to press.[30] A survey of its content permits the supposition that Pascal had also contributed to it. The chapters in question are devoted to cognition based on faith. Faith may be secular (*humaine*) or religious (*divine*). In both cases, we invoke certain arguments which – if the cognition is to be rational – must be properly 'balanced.' These analyses, invoking faith in the religious sense, established a certain model/standard repeated afterwards in different variants in numerous treatises and theological manuals (in what used to be known as apologetics, today called fundamental theology). Analyses invoking faith in the secular sense can be interpreted as precursors – as Hacking[31] says – of the later logic of not-deductive inferences, or those in which the certainty of results must be replaced with a certain degree of probability. Here appears a word which is crucial for our considerations – 'probability.' Hacking asks: When did this word appear for the first time to designate something 'that can be measured?'[32] In the correspondence between Pascal and Fermat, the word 'probability' in this sense was never used. It occurred right here, in the final chapters of *Port-Royal Logic*. In Chapter 14, we can read that in the event that we cannot entertain complete certainty in a given matter, and must take sides, we "should adopt most probable position, because to adopt a less probable one would be to

---

[30] Cf. A. Hacking, *The Emergence...*, *op. cit.*, p. 74.
[31] *Ibid.*, p. 75.
[32] *Ibid.*, p. 73.

negate rationality" (*un renversement de la raison*).[33] In many cases (legal, medical, or those in our daily lives) it is enough to invoke common sense and objective consideration of all circumstances, but there are situations in which we can offer a numerical measure of such probability. Who, for example, would not like to win a game of chance?

Let us imagine that ten individuals are playing such a game. Everyone contributes one écu[34] to the prize pot and the winner takes everything. Most people tend to believe that the game is worthwhile, because they may lose one euro at worst, and they may win ten, but one must take into consideration the fact that for every player, the probability of winning is nine times smaller than probability of losing, which completely balances out the fact that one lose one euro, and win nine. The game is fair, because every player has an equal chance of winning. The conclusion is as follows:

> In order to judge of what we ought to do in order to obtain a good and
> to avoid an evil, it is necessary to consider, not only the good and evil
> in itself, but also probability of its happening and not happening, and
> to regard geometrically the proportion which all these things have,
> taken together.[35]

This already constitutes a very distinct programme of attributing to different events their probabilities expressed as' geometrical proportions,' or, in other words, in terms of numerical measure. Here, one can perceive a distinct trace of Pascal's lion's claw.

The closing paragraphs of *Logic* appeal to the Reader that he in his life should not be guided by appearances, but encourage him to critically evaluate the probabilities of different events. The principle "there is danger in that business; therefore, it is bad: there is advan-

---

[33] *Logique de Port-Royal*, Paris 1854, pp. 319–320.
[34] The écu was an old French coin. It was also suggested as the name for the currency in the European Union (European Currency Unit), but the *euro* eventually won.
[35] *Logique de Port-Royal, op. cit.*, p. 319–320.

tage in this; therefore, it is good is delusive since "it is neither the danger nor the advantage, but the proportion between them, of which we are to judge."[36]

These proportions can be very sophisticated. It may happen that a very small gain will exceed a greater profit if the former occurs very often, whereas the latter only exceptionally.

This is the case among the finite things, whereas it is otherwise in matters concerning "infinite things, such as eternity and salvation," since they cannot be compared with any temporal advantage, none of the things of this world can equal them: "this is why the smallest degree of facility for the attainment of salvation is of higher value than all the blessings of the world put together."[37]

3. The wager

At this point, one cannot but mention Pascal's famous wager. Even though it is doubtful whether the last chapters of the *Port Royal Logic* were written by Pascal, it is known for a fact that he was the author of the wager. The wager is to be found in his *Thoughts* – a collection of reflections from scattered notes and compiled into a book. It is possible that some of its portions were intended for further elaboration and were to constitute parts of the planned *Apology of the Christian Religion*, Pascal never fulfilled this intention. The essence of his wager can be summarized as follows:

> I am facing a choice – either God exists, or He does not exist.
> If I choose 'God exists,' and God does not exist, I risk my finite life.
> If I choose 'God does not exist," and God exists, I risk an infinite eternity.
> So the choice is obvious.

---

[36] *Ibid.*, p. 324.
[37] *Ibid.*, p. 326.

It is a little known fact that Pascal had his own medieval predecessor. Crombie quotes the story of a clergyman living in the 12th century who declared on his deathbed that he would not believe in future resurrection if he had not been convinced by arguments based on probability. This clergyman was

> confident in his success given that if he was right, he would not have risked his faith and soul if he rose again and, if he was wrong, then he would never have found out about it.[38]

The original text of Pascal's wager is found on two handwritten pages and contains numerous crossings-out and corrections. Enormous interest in this text has meant that every spot of ink has been subjected to thorough analysis by researchers.[39] In the published versions of *Thoughts*, the wager can be found in a section entitled *Infinity – nothing?*[40]. The reasoning was recorded in a form far removed from linguistic accuracy. It looks like impromptu notes made on the spur of the moment and occasionally takes the form of a dialogue with an opponent. Let us now take a closer look at the wager.

First, the direct context. Pascal wonders about the relationship of finiteness to infinity and compares it to our relationship with God:

> Unity added to infinity increases it not, any more than a foot added to infinite space. What is finite, vanishes before that which is infinite, and becomes absolutely nothing. For instance, our understanding, in respect of God's...

---

[38] Cf. A. C. Crombie, *op. cit.*
[39] I. Hacking, *The Emergence...*, *op. cit.*, p. 63.
[40] Individual fragments are numbered, but their arrangement varies by edition. The author follows the numbering scheme used in the Chevalier edition. The wager appears in fragment 451.

If God is infinite, we "are unable to comprehend either what He is or whether He is." But Christians believe that God exists. So should their faith be considered a manifestation of stupidity (*stultitia*)? The text that follows must be an answer to this question, thus it is a kind of apologetics, an attempt to demonstrate that the Christian faith is not irrational. Pascal further reasons:

> Yet this remains certain, that either God is, or is not; there is no medium. Which alternative, then, shall we prefer? Reason, say you, is not a proper judge in this point. There is an infinite gulf between us. At this immense distance, we are playing, as you conceive, a game of chance. What then will you wager?

In a way, Pascal had already achieved his aim by demonstrating that the faith of Christians in God is not irrational, at any rate, no less rational than the faith of those who reject the existence of God ("Do not be forward, then, in accusing those of error, who have already made their choice. For you cannot be certain that they have acted imprudently, and chosen badly."). As is the case with the game of tossing a fair coin: heads or tails, fifty-fifty chance. But Pascal goes still deeper.

In the context of uncertainty, it is best not to make any choice at all ("the right course had been not too wager at all"). But this choice cannot be avoided: "Nay, but there is a necessity of wagering; your interest is embarked in this question". One is forced to wager in the sense that not to make such a choice would be irrational. Here the essence of Pascal's reasoning:

> Let us balance the gain and the loss. If you gain, you gain all; if you lose, what you lose is nothing.

And the inescapable conclusion:

> Believe, then, if you are able, without delay.

This reasoning has caused (and still causes) emotional reactions on both sides: some interpret it as effective apologetics,
others – as an abuse of logic for the sake of a religious propaganda. We will not enter into such disputes.[41] Pascal's reasoning can
be considered from several vantage points: (1) with special attention paid to its logical correctness, (2) as a model for decision
making, (3) as a strategy in the spirit of the game theory. Certainly, all these points of view are closely associated with one another. One cannot make a good decision if it is based on a logical error. Game theory must also respect logic. As I have already
mentioned, Pascal's writing is far removed from linguistic accuracy, still it can be reconstructed logically and thus ascertain its
correctness. Hacking[42] undertakes as many as three such recon-
structions (based on different decision-making models) and in all
three cases attests to the logical correctness of Pascal's reasoning. Nevertheless he thinks that Pascal's argument is not convinc-
ing, because it rests on "at least doubtful premises." He maintains
that no contemporary agnostic would change his convictions because of Pascal's argumentation. The most doubtful premise of
this argumentation is – in Hacking's view – the dichotomy: if you
do not believe in God, and God exists, you will be condemned vs.
if God exists and you believe in Him, you will be saved. Indeed,
challenging this premise undermines the conclusion of Pascal's
reasoning, but if anyone builds their agnosticism only on the basis of such a challenge, then I think Pascal may have achieved his
objective. As usual in such cases, a crucial role is played not by
formal-deductive fine points but by the existential power of the
argument.

---

[41] For a more extensive treatment of the subject, see e.g. S. Wszołek, *Trzy funkcje
zakładu w Myślach Pascala (Three functions of Pascal's wager in his Thoughts)*,
"Kwartalnik Filozoficzny" 2003, 31, pp. 83–100.
[42] I. Hacking, *The Emergence...*, *op. cit.*, pp. 64–65.

4. Huygens – how to expect winning?

Let us, however, return from the heights to which we have been led by Pascal, to considerably lower stakes. The next step on the way to contemporary probability calculus was made by Christian Huygens. During his stay in Paris in 1655, he familiarised himself with the results of Pascal and Fermat's work in the area of theory of games of chance and took a keen interest in these matters. Soon afterwards, at the behest of Frans Van Schooten, a mathematician and a publisher, he wrote in Dutch a treatise entitled *Van Rekeningh in Spelen van Geluck*.[43] After a preliminary agreement with Huygens, Van Schooten translated this work into Latin and published in 1657. The treatise is titled *De ratiociniis in ludo aleae.* The Dutch version of the work was not published until 1660.

A key notion in Huygens' analyses is the concept of expectation (*expectatio*). From the psychological point of view, such a positing of the issue is justified: a player who starts a game of chance, expects to win. However, Huygens was not concerned with subjective feelings of players, but with a notion that would permit him to introduce a quantitative element to predictions associated with these games. Today's notion of probability was not yet known (because it is hard to treat its mention in *Port Royal Logic* as common knowledge) and founding of calculations on *expectatio* constituted an important achievement. Today, the concept of expectation also appears in different contexts in probability calculus. For example, if we repeatedly roll a (fair) die, it is easy to calculate that the average result will be close to 3.5. This number is called the (weighted) expected value of repeated rolls of a die.[44] Our contemporary understanding of expectation should not be imposed on Huygens, the less so that Van Schooten was not very accurate when he translated the Dutch word *kansse*

---

[43] *Calculations in Games of Chance.*
[44] In the case of rolling a die, it is calculated using the formula: $1/6 + 2/6 + 3/6 + 4/6 + 5/6 + 6/6 = 3.5$.

into *sors seu expectatio*.[45] The word *expectatio* (which, in fact, we owe to Van Schooten, not to Huygens) in its sense intended by Huygens, is close to today's meaning of the expected value, but not identical to it. At the very beginning of his treatise, Huygens describes the following situation. Someone is engaged in a game of chance, but at a certain moment wishes to sell his place in the game to somebody else. How much should he demand? Huygens answered: he should demand as much as would be procured in the expectation of a win, and the expectation (*sors seu expectatio*) of a win equals the amount that can be expected in a fair bet. For example, somebody puts 3 shillings in one hand, and 7 shillings in the other, giving me a chance to choose 'blindly.' What chance do I stand or how much is my choice worth? Huygens answered: (3+7)/2 shillings, that is 5 shillings. In this case, the value equals our expected average value, but the concept is still entangled in a specific situation, deprived of a clear definition. Even though Van Schooten did not translate the Dutch expression used by Huygens very accurately, the inaccuracy was rather felicitous. The terminology took root and propelled further considerations concerning games of chance in the right direction.

---

[45] Cf. H. Freudenthal, *Huygens' Foundations of Probability*, "Historia Mathematica" 1980, 7, pp. 113–117.

Chapter 3

Theology versus probability

1. Posthumous work by Jacob Bernoulli

As fortune would have it, Huygens' short work had an impact dreamed of by all the authors of scientific publications: it became a link that secured the subsequent stages of development of important ideas. 1713 saw the publication of Jacob Bernoulli's posthumous work entitled *Ars conjectandi*.[46] In its first part, Bernoulli rewrote Huygens' work, adding numerous important comments, and in further parts he significantly expanded the entire theory. In *Ars conjectandi* for the first time the term 'probability' appeared in its technical sense, in which it has been used until this day. Moreover, thanks to this work, considerations on the games of chance turned into mathematical probability theory.[47]

In 1622, the Protestant family of Bernoulli arrived at Basel, a Calvinist city, seeking refuge. The young Jacob Bernoulli (born in 1654 in Basel), as was his father's wish, was to become a minister and read theology at the University of Basel. Although eventually he did not become a minister, he was fervently religious and maintained a keen interest in theology. During his journeys, he readily associated with theologians and Protestant intellectuals. His *Ars conjectandi*,

---

[46] Jacob Bernoulli, *The Art of Conjecturing*, together with a *Letter to a Friend on Sets in Court Tennis*, Basileae 1713. Translated into English by Edith Dudley Sylla, John Hopkins University Press, Baltimore 2006.

[47] Cf. G. Shafer, *The Significance of Jacob Bernoulli's Ars conjectandi for the Philosophy of Probability Today*, preprint.

which played such an important part in history of probability calculus, was – as can be supposed – the result of creative struggles between his theological views and the instinct of a mathematician. Bernoulli worked on it for twenty years. He died in 1705, without having published it. It was not until 1713 that the treatise was printed thanks to the endeavours of his nephew, Nicholas Bernoulli, in its original unfinished form.

The title *Ars conjectandi* clearly refers to the Latin title of *Port-Royal Logic, Ars cogitandi*, and, as Hacking says, the art of conjecture (*conjectandi*) takes up the issue where the art of thinking (*cogitandi*) has concluded its investigations.[48] The work consists of four parts. Part One – as I mentioned above – links directly to Huygens' achievements and develops them. In Part Two, Bernoulli builds combinatorics in order to apply it in Part Three to gaming problems. Part Four, the most important one, contains Bernoulli's main achievements: the first limit theorem of probability calculus, a discussion of the notion of probability itself and its applications to different areas of social life. We shall deal with them in turn.

2. God's omniscience and the law of large numbers

According to Bernoulli's theorem, which was later named by Poisson the law of large numbers, the average of the results obtained from a large number of trials (e.g. rolls of a die or tosses of a coin) approaches the expected value (and approaches it the closer, the more trials are performed).[49] The concept of expected value was mentioned in Chapter 2.4, where we quoted an example of expected value for rolls of a die. If the random variable is of a discrete type, i.e. it adopts discrete

---

[48] I. Hacking, *The Emergence...*, *op. cit.*, p. 145.

[49] Bernoulli proved his theorem only for binary random variables, in other words, for random experiments with two possible outcomes (success/failure); thus, a coin toss constitutes a Bernoulli trial, whereas a die roll does not.

values $x_1, x_2,..., x_n$ (in die rolls, these values equal 1, 2, 3, 4, 5, 6) with respective probabilities of $p_1, p_2,..., p_n$, the expected value W can be calculated using the following formula:

$$W = \frac{x_1 p_1 + x_2 p_2 + \cdots + x_n p_n}{p_1 + p_2 + \cdots + p_n}.$$

If, in the case of die rolls, we assume that $p_1 = p_2 = ... = p_n = 1/6$, we will obtain a formula from footnote 44 in Chapter 2.4, which yields 3.5 as the expected value. According to the law of large numbers applied to the process of die rolling, the average of results obtained from a large number of trials approaches 3.5 and the longer the sequence of rolls, the closer the average of results approaches this number.

Somewhat more generally, let s_n denote the number of successes obtained in n trials. $Z_n = s_n/n$ represents thus the average number of successes in n trials. Then, for every positive number ε there exists a parameter m such that for every $n > m$, the distance between s_n and W is smaller than ε. This is precisely what the law of large numbers is about.[50]

This law plays an important role in probability calculus. It warrants a certain kind of stability of random events. Thanks to it, the State Lottery always achieves substantial profits, although every now and then it must pay out a large sum of money to the lucky winner.

This law (or, to be more precise, its special case) was formulated and proved by Bernoulli in Part Four of his *Ars conjectandi* (Chapter 5), and in his proof used the combinatorial methods developed in the previous chapters. The impact of his achievement for the development of probability calculus is hard to overestimate. It steered the probabilistic considerations about gambling in the direction of a full mathematical theory.

---

[50] In its so-called weak form. There is also a strong form of the theorem. Both differ with respect to how the average 'approaches' the expected value.

However, we must not forget that when Bernoulli was formulating his theorem, his mind and creative imagination did not work in a milieu of ready-made formulae, ready to be manipulated (as is done today by numerous users of probability calculus), but struggled with a problem that had to be correctly formulated and solved. In such situations, assistance is sought from all possible sources, especially those to which the thought gravitates most readily.

Jacob Bernoulli faced the following situation: there was a sequence of events today called random events. I wish to find a mathematical formula which would help me to determine the probability of an event favourable for me (success), but this event lies in the future. How to do it if the future is undetermined? Bernoulli was aided by his Calvinist beliefs, among which faith in the predestination plays an important role. At the very beginning of Book Four, he writes:

> Everything that exists or originates under the sun, – the past, the present, or the future, – always has in itself and objectively the highest extent of certainty. This is clear with regard to events of the present or the past: because, just by their existence or past existence, they cannot be non-existing or not having existed previously. Neither can you have doubts about (the events of) the future, which, likewise, on the strength of Divine foresight or predetermination, if not in accord with some inevitable necessity, cannot fail to occur in the future.[51]

Accordingly, the possibility of formulating probability-based predictions of events is justified in the determination of the future – if not a determination inherent nature itself, then at any rate thanks to God's omniscience or predestination. But this objective determination corresponds with, on our side, our ignorance about the future. And here, for the first time ever Bernoulli formulated an

---

[51] *Ars conjectandi, op. cit.,* pp. 210–211.

important distinction for the interpretation of probability – into objective and subjective certainty. Objective certainty "does not mean anything but the truth itself of things existing or future," however, subjective certainty consists in "the measure of our knowledge of this existence."[52]

The future is determined objectively, which guarantees the possibility of predicting events, but subjectively, the future remains unknown to us, therefore in its prediction one should apply methods of probability calculus.

So the law of large numbers finds its philosophical (ideological) consolidation – all the arguments add up to form a harmonious whole. Yet Bernoulli perceives in it a theological difficulty: How to reconcile the objective determination of phenomena (on the strength of Creator's omniscience) with the contingency of the world and "the freedom of secondary causes?"

> Let others argue about it – writes Bernoulli succinctly – we, however, will not touch something alien to our aims.[53]

Indeed, in the course of his work Bernoulli does not revisit this issue, but – as we will see – there are serious arguments suggesting that when he was writing his *Ars conjectandi*, these theological problems still constituted the focus of his interest and inspiration.

For Bernoulli, the law of large numbers had a theological as well as a practical significance. Thanks to its generality, it demonstrated that strategies of the numerical prediction of results could be transferred from gambling onto other areas of human activity. Part Four of Bernoulli's work is devoted to "The Use and Application of the Previous Doctrine to Civil, Moral and Economic Affairs." Gambling was due to become, over time, only a historical problem and a source of numerous examples discussed in probability calculus.

---

[52] *Ibid.*, p. 210.
[53] *Ibid.*, p. 211.

3. Probability

In order to denote mathematical analyses, which gradually began to surround the speculations on gambling games, the term 'gambling geometry' (introduced by Pascal) was applied. Interestingly, in the first works in the area of 'gambling geometry' the term 'probability' was not used. As we remember (Chapter 2.2), the term was used for the first time in Chapter 14 of *Port-Royal Logic*, but it was not sufficient for it to become a technical term. To some extent, the role of today's probability was fulfilled by the term 'expectation' (*expectatio*) introduced by Huygens, it turns out, however, that the notion of probability was extensively applied in an area quite unrelated to 'gambling geometry,' namely in moral theology, mainly by the so-called Spanish probabilists. Moral probabilism held that in the case of uncertainty as to how one should proceed, one should follow a probable opinion. For the most part, it was applied to situations in which such decisions could not be postponed, but the area of interest also included more theoretical problems: How to apply general moral norms in a situation that gives rise to doubt as to which moral precept should be followed? This doctrine was for the first time explicitly formulated by Bartolomé de Medina. In his view, an opinion is probable as long as it is supported by substantial arguments and is shared by "learned men." In this approach, probability is the property of an opinion or an utterance that expresses such an opinion.

The term 'probability' was used by Jacob Bernoulli in Part One of *Ars conjectandi*, in his commentary to Huygens paper *De ratiociniis in ludo aleae*. Furthermore, somewhat earlier, as a conclusion from Huygens' Propositio III, Bernoulli derives a definition of probability (though he does not call it thus yet) as the proportion of the number of favourable outcomes to the number of all possible outcomes.[54]

---

[54] *Ibid.*, p. 9. Cf. J. Santos del Cerro, *L'ars conjectandi. La géométrie du hasard versus le probabilism moral*, "Journal Électronique d'Histoire des Probabilités et de la Statistique/ Electronic Journal for the History of Probability and Statistics" 2006, vol. 2, no. 1, pp. 1–23.

Bernoulli introduces the notion of probability systematically at the beginning of Part Four and he opposes it to the concept of certainty. In accordance with the tenets of his philosophy, Bernoulli stipulates subjective certainty ("in the relation to us"), which may be gradable, "another is about which we know, either by revelation, intellect, perception, by experience, autopsia [direct observation; by one's own eyes] or otherwise.[55] On the other hand, probability is "the degree of certainty." Bernoulli immediately establishes a convention: if "absolute certainty" is denoted by 1 (one), then "something is possible if it has 1/20 or 1/30 of certainty."[56]

Seemingly, the idea of epistemological (subjective) probability does not have a lot in common with the issue of chance in gambling games, but Bernoulli applied Huygens' method of expectations (and even the same mathematical formula) in order to represent probabilities in numerical terms. To the same end, he applied his own law of large numbers, namely, as a tool for determining "how much more easily can some of them occur than the others."[57]

In spite of its subjectivist position, Bernoulli noticed the difference between random events in the games of chance and those that are "the production of nature,"[58] and also those which originate in man's free will. In gambling, we may know the relative frequency of occurrence of simple outcomes and we are entitled to thinking that the same frequency will occur in the future, but in the case of undetermined nature and man's free will, we have no such option, we do not even know for certain whether we have the right to follow the analogy with gambling. What is left to do is to determine probabilities *a posteriori* on the basis of past experience.

Bernoulli's preliminary considerations reveal traces of his inspiration: he calls a conviction "whose probability is almost equal to complete certainty" morally certain ("so that the difference is in-

[55] *Ars conjectandi, op. cit.*, p. 211.
[56] *Ibid.*
[57] G. Shafer, *Non-Additive Probability in the Work of Bernoulli and Lambert*, "Archive for History of Exact Sciences" 1978, 19, p. 309–370.
[58] J. Santos del Cerro, *op. cit.*, p. 14.

sensible"). Shafer[59] argues that Bernoulli's great achievement was to perceive the relationship between probability in games of chance and probability as disputed by moral philosophy and theology.

4. Moral disputes

Although Bernoulli promised not to get involved in theological disputes, and, for the most part, he kept his word, commentators tend to agree that he harboured a deep concern for theological problems and inspired a number of his scientific ideas; consequently, it makes sense to look in somewhat more detail at the theological context of Bernoulli's views.[60]

The relation between God's grace and man's free will was a serious theological issue dating back at least to the times of St. Augustine of Hippo. In the middle of the 16th century, disputes on this subject flared up with renewed vigour fuelled further by controversies caused by the Counter-Reformation. The dispute revolved around the question of whether God, knowing the future, can prevent moral evil, granting in advance appropriate grace to enable man to overcome temptation. The Dominican school, headed by Domingo Bañez, adhered to the traditional doctrine. God gives man freedom to decide and grants him sufficient grace, though not always effective, to overcome the temptation. God knows the future by insight and from His extratemporal perspective simply sees what man will do without interfering with his freedom. Luis de Molina, a Jesuit who began the entire dispute with his work (called *Concordia*, for short), took a different approach to this problem. In his opinion, God's cooperation with man is 'neutral' and it depends on man whether the grace that he receives will be effective or only sufficient, yet unused. God's knowl-

---

[59] G. Shafer, *op. cit.*
[60] More information on the subject can be found in J. Santos del Cerro, *op. cit.*

edge of the future is in a sense constrained. Molina called it *scientia media*. The knowledge lies, as it were, in between the awareness of all possibilities and what will actually happen.

The dispute was so fierce that Pope Clement VIII saw fit to appoint a select commission to investigate the matter. The theological difficulties inherent in the dispute are perhaps best reflected by the fact that the commission ruled in 1607 that both sides might continue to defend their respective positions.

It is not hard to infer that Bernoulli's position was closer to that held by the Dominicans, because it was easier to reconcile with the Protestant doctrine of predestination. In an attempt to determine the frequency of future events, Bernoulli – in keeping with his theorem of large numbers – used combinatorics to put together all the possibilities that may occur. Such an approach can be interpreted as an attempt to look at a process that occurs in time, as it were, from the perspective of God, who takes in the past and the future with a single look, thus mathematics becomes in a sense such a 'God's perspective.' Conversely, granting God middle knowledge (*scientia media*) after Molina, would rule out such an interpretation.

5. Bernoulli and Leibniz. The logic of contingent events

In 1703–1705, Bernoulli and Leibniz exchanged at least 20 letters, many of them devoted to issues of probability. Leibniz was unable to familiarize himself with the manuscript of Bernoulli's treatise, he only knew its contents second-hand, yet at once he understood the importance of Bernoulli's analyses to the problems of formalisation of certain aspects of policy and legal matters that he himself found of interest. Bernoulli's theological motivation also attracted Leibniz's attention.

Bernoulli believed (on the basis of his law of large numbers) that by investigating the frequency of random events empirically, one could arrive at a "perfect estimation" of their occurrence, but Leibniz thought that given the contingent nature of the world, events that occur in it depend on an infinite number of circumstances, which cannot be determined by investigating only their finite number. In compliance with Leibniz's philosophy, e.g. the fact that Brutus killed Caesar, could be deduced from the "first principles," but the sequence of deductions that would lead from the "first principles" to a contingent fact, which was the killing of Caesar, was infinite and could not be captured in any "algebraic formula," which Bernoulli appeared to allow. Only God is in a position to know the conclusion of such an infinite sequence of deductions, because, existing outside time, He grasps everything "with His own eyes."

In Chapter Two of his *Ars conjectandi*, Bernoulli analyses the meaning of the term used in the title of his work.

> To make conjectures about something (*conjicere*) – Bernoulli writes – is the same as to measure its probability.[61]

He goes on:

> Therefore, the art of *conjecturing* or *stochastics* (*ars conjectandi sive stochastice*) is defined as the art of measuring probability of things as exactly as possible, to be able always to choose what will be found the best, the more satisfactory, serene and reasonable for our judgements and actions. This alone supports all the wisdom of the philosopher and the prudence of the politician.[62]

Further on in this part, Bernoulli offered not only a number of remarks, recommendations, but also numerical estimations of probabilities related to many situations in which we must make do with incomplete knowledge, such as in judicial, medical, social and ethical cases. This part of Bernoulli's reasoning aroused

---

[61] *Ars conjectandi, op. cit.*, p. 213.
[62] *Ibid.*

Leibniz's keen interest, since he himself had been for a long time fascinated by the problem of formalising legal procedures. He re-alised that problems related to the weight of judicial evidence in principle did not differ from the issue of evaluation of arguments based on incomplete proofs as debated in epistemology. If logical arguments can be applied to the latter, the same should be the case with the former. Even in his early treatise *De conditionibus* Leib-niz tried to apply the rules of logic to legal situations that entailed a decision whether someone was entitled to a certain right granted to him under certain conditions, when it was not obvious to what extent these conditions had been met. In his considerations, Leib-niz came closest to probability calculus when working on his *Ars combinatoria*; nevertheless, he did not make the decisive step of harnessing combinatorics to serve probabilistic calculations. As Hacking remarks,[63] admittedly Leibniz did not develop probabil-ity calculus but he contributed significantly to its development through the 'conceptualisation' of numerous related problems, i.e. by 'positing' certain concepts in a way that later facilitated their use in a mathematical context.

In Part Four of his *Ars conjectandi*, Bernoulli initiated the pro-cess of applying probabilistic considerations (which afterwards led to the establishment of statistics) to various manifestations of public life. Leibniz, having no suitable probabilistic tools at his disposal, even objecting to their use in the treatment of 'contin-gent events' – chose another way in that he attempted to subdue a wealth of real-life situations to the order of logical methods. And it was not only in theory. When necessary, Leibniz did not hesi-tate to apply mathematical methods to analyses of genuine con-tingent events. Moreover, he maintained that conclusions that he had drawn were certain. The context of this achievement was as follows.

---

[63] I. Hacking, *The Emergence...*, *op. cit.*, p. 89.

After the abdication of Jan Kazimierz (John Casimir), king of Poland, on September 16, 1668, a struggle for succession ensued amongst a dozen or so contenders. One of them was Philip William, Count Palatine of Neuburg. Leibniz, who was then in his service, joined the pre-election campaign on behalf of his patron. In 1669, he wrote a treatise titled *Specimen demonstrationum politicarum pro elegendo rege Polonorum novo scribendi genere ad claram certitudinem exactum*[64], in which he mathematically proved that Philip William was the best candidate for the Polish throne. Doubtless, in order to impress the reader even more, the pamphlet was indeed fashioned like a mathematical treaty: it comprised 60 propositions (*propositiones*), 12 supplements and 4 conclusions. The expression *specimen* (understood here as an example or model) in the title indicates, however, that Leibniz also had a more doctrinal aim than only propaganda in the interest of his candidate. He wanted to demonstrate that there existed "mathematically measurable" criteria of decision-making. In this way, politics contributed to the maturation of scientific ideas.

6. Political arithmetics

As we have seen, in Bernoulli's work probability calculus was sufficiently developed to warrant a practical application, however, he lacked the empirical data that would permit him to fully exploit the opportunities offered by the mathematical apparatus that he had invented.

Probability naturally leads to statistical analyses. The term 'statistics' appeared earlier in Girolamo Ghalini's work *Annali di Alessandria*[65], which dealt with *civile politica, statistica e milita-*

---

[64] *An example of political proof related to the election of the king of Poles conducted in a new mannner of writing in order to accomplish definite certainty.*

[65] *Annali di Alessandria, overo le cose accadute in essa citta nel suo e circonvivino territorio dall'anno dell'origine sua sino al MDCLIX*, Milano 1666.

re scienza. Only somewhat later did the meaning of the term shift from purely descriptive aspects to numerical analyses involving probability. The term 'political arithmetics' was also used.[66]

The systematic collection of statistical data began in London in 1603 following a major outbreak of the plague, and in Holland in connection with annuities[67]. These data came to be used for practical purposes: a search for the causes of the plague, forecasting the occurrence of a given phenomenon (e.g. male and female births), the profitability of a given type of annuity etc.. Especially eminent figures in this area were John Grant and William Petty. Grant formulated a principle on which all such research was based. To wit, he observed that statistical regularities occur with a large number of cases, yet remain invisible when their number is small. Incidentally, it may be interesting to note that Edmund Halley developed a special liking for Wroclaw as a city which was especially conducive for statistical research. In his opinion, it was a city so isolated that all who died in it had also been born there, and therefore it was a very convenient area to collect valuable statistical samples.[68]

This kind of research gradually acquired greater significance with statistical analyses finding an ever-broadening range of applications. This fact had a crucial impact on the further development of probabilistic methods.[69]

---

[66] Cf. A. C. Crombie, *op. cit.*

[67] What I understand by an annuity is the following transaction: X transfers to Y a certain amount of money that Y repays for a certain period of time. This kind of annuity may be temporary, if the transaction comes to an end once Y has repaid the whole amount, or perpetual (indefinite), if Y's repayments continue until the end of his life.

[68] Cf. A. C. Crombie, *op. cit.*

[69] A more extensive treatment of these issues can be found in Ian Hacking's *The Taming of Probability*, Cambridge University Press, Cambridge 1990, and *The Emergence...*, *op. cit.*

Chapter 4

Chance and how the world is organised

1. The principle of least action

The gradual taming of chance by mathematics and the application of probabilistic reasoning to public and economic phenomena could not have gone unnoticed by the philosophy of nature. Aristotle's view, that by breaking causality chance destroys its rationality, was too influential, therefore the recognition of the operation of chance in the world had to call into question the belief in a rational Creator of the Universe. An interesting attempt to go beyond the disjunction – either the Rational Creator or the operation of chance – was undertaken by Pierre-Louis Moreau de Maupertuis in his principle of least action. Initial intuitions can be sought in Leibniz's doctrine that our world was chosen by the Creator, because it was the best of all possible ones. According to Planck's all too optimistic diagnosis, "this theorem is none other than a variations principle, and is, indeed, of the same form as the later principle of least action."[70] Thus, we have a family of all possible worlds (since it remains unknown whether it can be called a set in the technical sense of the term). Put together, they comprise a collection of all possible, and hence random, situations[71], but the tenet of 'optimum choice' makes this random collection an intri-

---

[70] M. Planck, *The Principle of Least Action*, in *A Survey of Physical Theory* (formerly titled: *A Survey of Physics*), translated by R. Jones and D. H. Williams, Dover Publications, New York 2011, p. 200).

[71] In the sense that every randomly occurring situation belongs to the collection.

cately organised project and testifies to the rationality of the Creator who followed such a rational principle.

The same idea can be applied to random events in our Universe, but first it is necessary to find an appropriate principle that would optimise what really happens. This is exactly what Maupertuis did. In our Universe, phenomena occur in a way that minimises the quantity called action. Its prototype was Fermat's observation made in 1662 (also called Fermat's principle) that the path taken by a ray of light between point A and point B is the path that can be traversed in the least time, but only Maupertuis reformulated this observation and raised it to the rank of a rule that applied to all phenomena in nature. This rule became almost his life's obsession and he staunchly defended its universal character and his own claim to primacy in its discovery.

Nonetheless, the thought inherent in the principle of least action remains hazy until one defines the quantity which is supposed to be minimised, namely action itself. To Maupertuis' undoubted credit, this is what he tried to define. In his view, action was a product of the mass of a given body, the distance covered by it and its velocity. As history was later to show, this definition had not yet been accurate.

Maupertuis presented his conception for the first time in 1741 to the Paris Academy of Sciences. Three years later he elaborated upon it, this time for the Berlin Academy of Sciences. The title of his presentation testifies to the importance that he attributed to his principle: *The Conformity of many laws of nature which hitherto seemed incompatible* (*Accord de plusieurs lois naturelles qui avaient parujusqu'ici incompatibles*). In his next major address, in 1746, again to the Berlin Academy, Maupertuis extended his rule from rigid bodies to point masses. Besides, he took every opportunity to popularise this principle, and in his subsequent work *Essai de cosmologie* of 1750, drew certain conclusions concerning the worldview and cosmology from it.

2. The polemics of an academic

Pierre-Louis Moreau de Maupertuis came from a moderately wealthy family of merchants. After a three-year term of service in the cavalry, he discovered his true passion – mathematics. He practised it in Basel under the tutelage of Johann Bernoulli, younger brother of Jacob (the author of *Ars conjectandi*). His first important papers dealt with Newtonian mechanics, which he disseminated in France, which was still very much under the influence of Cartesianism. Applying Newton's theory, Maupertuis calculated that the Earth was a flattened ellipsoid, whereas Jacques Cassini, using astronomical measurements, maintained that it was an elongated ellipsoid. The issue of the shape of the Earth was perhaps *the* hot topic of the day. The Paris Academy dispatched an expedition to Peru to perform the necessary geodetic measurements. A year later, another expedition was organised, this time to Lapland, headed by Maupertuis. Despite various difficulties and adventures, including the wrecking of his ship in the Baltic Sea on the way back (fortunately, the materials gathered during the expedition were salvaged), the journey proved to be successful. The measurements performed confirmed that the Earth was indeed a flattened ellipsoid.

Thanks to this accomplishment, Maupertuis became famous but soon his difficult character made itself felt. He attacked Jacques Cassini so harshly that even his friends were shocked. His relations with Johann Bernoulli became strained, but he grew closer to Voltaire.

When Frederick the Great, King of Prussia, decided to found an Academy of Science in Berlin, he proposed that Maupertuis become its president. Maupertuis accepted the offer and discharged this function for eight years. Euler was also a member of the Academy, with whom Maupertuis formed a closer relationship. Meanwhile his relations with the Paris Academy became tense and, for some time at Cassini's behest, he was deprived of its membership. It was only after Cassini's death in 1756 that his membership rights were restored and he was granted a lifetime salary.

3. How does the Universe work?

It is almost a rule in the history of science that if a researcher makes an important discovery (at least in his own eyes), he begins to look at all the natural phenomena from the angle of this discovery. Not infrequently such a point of view turns into a comprehensive philosophical vision. Jean-Pierre de Maupertuis and his principle of least action constitute a case in point. He believed that in this principle he had found an idea that united all the laws of physics. In his *Essai de cosmologie*, he expressed an opinion that it constitutes in a way an mechanism that underlies the operation of the entire Universe, which always chooses – anthropomorphically speaking – the route of the least 'effort;' it is thus a peculiarly understood principle of economy.

In his earlier works, Maupertuis also dealt with biology. He wrote, among other things, quite a well-known book titled *Venus physique*, devoted in part to the development of the human foetus. As the title indicates, he did not avoid – apart from strictly biological themes – certain erotic tones. His earlier achievements in biology involve the formulation of a hypothesis concerning the existence 'heredity units' that originate both from the father and from the mother, pointed out not only the variability of individuals, but also entire populations (naturally, I am using today's terminology).

With his own considerations in the domain of biology as a starting point, Maupertuis sharply argued against 'the teleological argument,' as it was understood in his day. In *Venus physique*, he wrote:

Chance, one may say, created an innumerable plurality of individuals; their small number was built so that body parts of the animal were able to serve the satisfaction of its needs, in another, infinitely larger number of individuals, there was neither an adaptation nor chance. All of them became extinct; animals without the mouth were unable to live, others, having no reproductive organs, could not survive. Only those remained in which order and adaptation appeared, and the species that we see today, make up only a minuscule part of those created by blind chance.[72]

Can we consider the author of these words as the precursor of the principle of natural selection? Opinions of historians of science are divided on this issue. The idea was doubtless floated, but was neither developed more fully nor underpinned by empirical material. Moreover, the above-mentioned quotation proves that it is necessary to employ quite intricate argumentation in order to defend Maupertuis from an accusation of a naive understanding of the operation of chance in natural selection. He understood it in a manner similar to the ancient atomists – we have all randomly put together possibilities: organisms without a mouth, without reproductive organs etc. and it is from this collection that selection chooses only those capable of surviving. However, we must admit that some of his statements are quite striking. For example, in *Systeme de la nature* he writes:

> Could we not explain by this, how, originating from only two individuals, a multiplication of all kinds of different species might occur? They would have owed their earliest beginnings only to some random creatures in which elementary particles did not keep such an order as they did in the father or in the mother of the animal; every degree of error would have yielded a new species; and through repeated deviations there would arise an infinite variety of animals which we se today; and which may still increase in the course of time or which may, in the course of centuries increase only imperceptibly.[73]

---

[72] Maupertuis, *Vénus physique*, retrieved at http://fr.wikisource.org/wiki/Vénus_physique.
[73] Maupertuis, *Système de la nature*, retrieved at http://books.google.pl/books/about/Systeme_de_la_nature.html.

Maupertuis' mathematical mind noticed that the order existing in nature could be explained by the mathematical game of probabilities without reference to an intelligent Creator. But the problem returned at a higher level, so to say. If the Universe always chooses the path of least action, then it is necessary to admit – reasoned Maupertuis – the existence of a Great Intention behind it. Maupertuis reasoning was as follows:

1. From the principle of least action, one can derive all the laws governing the behaviour of phenomena in nature.
2. These phenomena are exactly what we observe; they include, among other things: "the moving of animals, the vegetation of plants or the revolutions of the heavenly bodies."

In conclusion:

> These laws, so beautiful and yet so simple at the same time, are maybe the only ones that the Creator and Ruler of all things established for matter in order to make it possible for all phenomena of the visible world to occur.[74]

In this way, the previous 'teleological argument' (design argument) was replaced by a new one 'derived from the principle of least action.' Purposefulness cannot be perceived at the level of individual phenomena, because all can be explained by probability and selection. The Creator's thought can be deciphered only from the principle on which the laws of nature are founded. This is an example of treating chance as a constituent 'of a design at a higher level.'

---

[74] Maupertuis, *Essay de Cosmologie*, retrieved at http://fr.wikisource.org/wiki/Les_Œuvres_de_M._de_Maupertuis.

4. The case of Samuel König

In the history of science, great ideas are interleaved with instances of human weakness. Maupertuis 'practised mathematics' under the tutelage of Johann Bernoulli together with Samuel König. Later, as president of the Berlin Academy, he proposed König's candidacy for its membership. Soon König submitted to the Academy an article for publication. Maupertuis, most likely not having read it, approved it for printing. Only afterwards did he realise that the article contained a fierce attack on his principle of least action. Moreover, the author credited Leibniz with the formulation of this principle, invoking a letter allegedly written by the scholar in 1707 to Jacob Hermann. As expected, Maupertuis reacted violently by, first of all, demanding to see Leibniz's letter. Some time later König managed to submit only a copy of the said letter, maintaining that it was passed on to him by Henzi, who had been beheaded three years before in Berne for attempted mutiny. At Frederick the Great's behest – since the honour of his Academy was at stake – a search was started, including the Berne archives. The letter was not found.

Euler stood up in Maupertuis defence. He wrote a long letter to the Academy in which he argued the case from different points of view, however, it is hard to interpret his stance as entirely impartial[75] but, owing to Euler's considerable authority and great knowledge, it is worth quoting several sentences concerning the principle of least action itself.

The letter starts with a statement:

> Mr. de Maupertuis, President of the Royal Academy, has shown (by several very convincing arguments) that the action is always minimised, not only at equilibrium but also in the motions of bodies under external forces; this remarkable *principle of least action* expresses the most general law of Nature.

---

[75] The text of Euler's letter can be found in en.wikisource.org/wiki/Investigation_of_the_letter_of_Leibniz.

And further:

It is as though Nature wishes to achieve some effect, and approaches
it as closely as possible.

The last sentence constitutes a reply to König's accusation that
quite often the action observed does not comply with the principle.
Indeed, emphasised Euler, the least value of action is not achieved if
it is prevented by constraints, whereupon Nature approaches the intended result as closely as it is only possible.

In conclusion, the assembled members of the Academy supported Euler's expert opinion and decided

that the passage published by Mr. König in the *Acta Eruditorum* of
Leipzig as a quotation from a letter of Mr. Leibniz written in French
to Mr. Hermann, appears to be a forgery and, consequently, does not
carry even a shadow of authority concerning the legitimate rights of
the Academy members involved in this affair (...)[76]

However, such matters are not forgotten quickly. A hundred and
fifty years later, the Berlin Academy returned to the issue raised by
König. With the benefit of hindsight, when personal emotions had
long since expired, a more matter-of-fact examination was undertak-
en. As a result, the authenticity of quotations from Leibniz's letter
was confirmed, but it was not until 1913 that it turned out that Henzi
had had at his disposal only copies of Leibniz's letter. These were not
the only copies. Others were found in the possession of Johann Bernoulli. Among them were the letters quoted by König.

Reading the quotation from Leibniz' letter referred to by König
with an unprejudiced eye, which can be found in Euler's address, one
may notice a detail that may be viewed as an argument in favour of
its authenticity. It starts with the following remarks:

---

[76] *Ibid.*

> The action is not at all what you imagine, since time is involved. The action is rather the product of the mass and the time, or the time multiplied by the kinetic energy. I have noticed that, in changes of motion, the action is either minimised or maximised. One can derive from that several important results (...)

In view of the fact that Leibniz defined kinetic energy as the product of mass and velocity, action as understood by Leibniz did not significantly differ from action in Maupertuis' sense. But neither Maupertuis nor his defender Euler noticed that in the course of events in nature activity is either minimised or maximised. They invariably spoke only about the former and it is difficult to imagine that this fact was noticed by König. At any rate, if he had, he would not have failed to use it in his dispute with Maupertuis. Leibniz understood what today can be found in every physics coursebook. The principle of least action is, in fact, the extremum principle.[77] If at present we mention the principle of least action, it occurs chiefly for historic reasons.

5. The principle of least action in contemporary physics

Did Maupertuis really make such a groundbreaking discovery? Looking from today's perspective, one might say that he was on the right track, but if he had devoted more energy to obtaining a deeper mathematical understanding of the problem than to advertising his achievement, he would have been able to achieve much more. Later, this principle was explored by Lagrange, Jacobi, Hamilton and others using considerably more accurate methods. Nonetheless, the principle of least action was considered for some time as a curiosity of sorts, devoid of practical consequences for the development of mechanics. The situation started to change when Bolzmann and Clausius found a close link be-

---

[77] In the language of mathematics, the concept of extremum refers to both minimum and maximum values.

tween the second law of thermodynamics and the principle of least action, with Helmholz systematically examining the operation of the principle of least action not only in mechanics, but also in thermodynamics and electrodynamics. Much later, Hilbert achieved a great success in deriving field equations of general relativity from the principle of least action. Today, we know that this is one of most fundamental principles of all physics, however, the name 'principle of least action' now has only a historical meaning, since – as physics progressed – it turned out that in numerous situations action is not minimised but maximised, so we should be talking – and sometimes even do so – about the principle of extremal action (or the extremum principle, for short).

Let us look somewhat more closely at the role played by the principle of least action in contemporary physics.

The work of Newton – the founding of classical mechanics – is monumental, but sometimes leaves us with a sense of being lost – a fat volume that must be studied, dozens of equations which one must plough one's way through... The magnitude of Newton's achievement also consists in the fact that it admits of a synthetic, unusually elegant formulation. Let us consider a system of mass points (material bodies) amongst which different forces operate. In order to 'describe' the dynamics of such a system, it is necessary to formulate a series of equations (according to Newton's prescription) and solve all of them. But the thing about mathematics is that quite often it permits us to replace a picture that is complicated and full of details with an aesthetically simple structure without prejudice to any of the details. Let us consider all the possible configurations of mass points within the system that we wish to examine. These configurations make up a space S called the configuration space – with every specific configuration constituting a point in this space. Let us choose a certain point P (a certain configuration of mass points) in the configuration space S, with which we wish to start our investigation of the dynamics of the entire system. In time, under the influence of various forces, the configuration changes, so point P marks a curve C in space S. This curve represents the history of the dynamically changing system. Yet curve C does not change in an unconstrained manner – it does so in a way designated by Newton's laws of dynamics. And here follows something that one might like to call a miracle of the method. It turns out that these laws are encoded in a single function[78] called Lagrange's function, or the Lagrangian. This function incorporates complete information about all that is necessary to determine the dynamics of a system. We disentangle it in accordance with a strictly determined procedure. To describe it in the most general terms, first we obtain a certain integral of the Lagrangian, which we call action. Then we compute the value of this action along all the curves in the configuration space that connect the initial and the final points of the dynamic process under investigation. This curve determines a dynamics that agrees with Newton's laws along which action has an extreme value, i.e. either is greatest or least.

Roger Newton thus describes the operation of the principle of least action in classical mechanics in the following way. In this approach, we use only a single function that contains all the information on forces, masses etc., and then we formulate a question: what should the motion of bodies or particles look like, starting with a given configuration in the initial moment to a given configuration in the final moment, so that the function, called action, has the least possible value? The motion minimising action fulfils Newton's equations, and this fact bears the name of the principle of least action or Hamilton's principle.[79]

6. Is the integral more mysterious than the differential?

What remains to be discussed is the philosophical problem that fascinated Maupertuis so much: Does the principle of least action really entail purposefulness? Quite often, a negative answer to this ques-

---

[78] Determined on a tangent bundle of the configuration space. More information on this subject can be found in R. Penrose, *The Road to Reality: A Complete Guide to the Laws of the Universe*, Vintage Books, New York 2004, Chapter 20.

[79] R. G. Newton, *What Makes Nature Tick*, Harvard University Press, Cambridge, MA 1993, p. 192. An exact formulation of Hamilton's principle with interesting examples can be found e.g. in V. I. Arnold, *Mathematical Methods of Classical Mechanics*, Springer, Berlin 1989, Chapter 3.13.

tion is supported by drawing attention to the fact that, once the rule is formulated properly, action is not so much minimised as either minimised or maximised. This argument is evidently ineffectual since the aim can equally well be to achieve minimum action as well as extreme action. Only thinking anthropomorphically we would like nature to strive for the least 'effort.' The integral of the Lagrangian is often represented as S, so the extremum principle can be expressed as $\delta S = 0$, which means that the variation of S equals zero. But the principle thus expressed can be read not as 'the principle of the minimum,' but as 'the principle of the optimum,' which, in turn, may well pass for the crucial principle adopted by the Designer of the World. Yet such an interpretation would be, in fact, an instance of overinterpretation, i.e. an interpretation that is not required by the mathematical formalism of the extremum principle, but imposed on it from outside. Let us look at it a bit more closely.

The formulations of the least action principle (let us stick to the inappropriate, but historically legitimised term) fall into differential and integral ones. The differential principles – including d'Alembert's principle and Bernoulli's principle – capture motion only in terms of local quantities, i.e. they construct motion from quantities that determine it in subsequent infinitesimally small neighbourhoods (e.g. from a change of momentary velocities). This shows no associations with purposefulness.[80]

The issue looks different in the case of integral formulations, which include, among others, Euler-Lagrange's principle and Jacobi's principle. The fact that in these formulations motion is expressed in terms of integrals, that is to say, quantities stretching over a longer time interval (or another integrated quantity), suggests that the future might be determined by what is occurring now, yet such an association is based on superficial intuitions that draw on anthropomorphist imagery. It is important that in order to step beyond these intuitions,

---

[80] More information on the subject can be found in D. Domaciuk, *Zasady wariacyjne i ich teleologiczna interpretacja (Variational Principles and Their Teleological Interpretation, in Polish)*, "Zagadnienia Filozoficzne w Nauce" 2008, 42, pp. 52–67.

we need nothing more but a rudimentary awareness of the relationship between the differential and the integral. Integration consists in 'adding infinitesimally small quantities,' so even here the global characterisation of motion is not formed by reaching straight to the end of the motion (to the future), instead, motion is constructed step by step from ensuing local quantities.[81] The concept of the integral does not conceal any more amazing mystery than does the concept of the differential.

Such an interpretation was advocated by many physicists, among others by Ernst Mach, who emphasised that from an analytical point of view, integral formulations were equivalent to differential ones. In keeping with J. Petzoldt's views, he underscored that the principle of extremum of action is nothing but an analytic expression of an experimental fact that phenomena in nature are unequivocally determined.[82] This should be understood in the sense that if nature is deterministic, it does not matter at all whether the evolution of a given system is computed on the basis of its initial or its final data. The entire behaviour of a system is unequivocally determined by its 'initial data' at any point of time. Yet there were physicists, e.g. Helmholtz and Planck, who maintained that integral formulations were more basic than differential ones. Moreover, the argument appealing to determinism loses its power if the principle of extreme action is applied to quantum theories, which, by their very nature, are indeterministic.[83]

7. Chance versus Intelligence

We remember the context in which Maupertuis formulated his idea of the Supreme Intelligence that manifests itself in the principle of least action. He noticed that "accidental combinations of

[81] Cf. especially pp. 62–64.
[82] E. Mach, *The Science of Mechanics*, Open Court, La Salle 1974, p. 471.
[83] A more in-depth treatment of the issue can be found in J.D. Barrow, F.J. Tipler, The Anthropic Cosmological Principle, Clarendon Press, Oxford 1986, pp. 148–153.

products of nature" plus the principle of advantage (adaptive selection) could explain the richness and harmony of nature. Thus, "a blind and necessary mechanics follows the designs of the most enlightened and most free Intelligence."[84] The greatness of this Intelligence – in Maupertuis' view – is reflected in the fact that it preferred to apply a single universal rule (of least action) to all creation rather than interfere with the emergence of every being or species separately. Maupertuis thus solved the problem of chance and the appearance of the new.

It remains unknown whether Darwin, working on his theory of evolution, knew Maupertuis' conception, but later (in 1868) he made at least one reference to it. In *The Variations of Animals and Plants under Domestication*[85] he drew a concise comparison of his theory of the origin of species through natural selection with the conception of "independent creations," every one of which should be treated "as the final fact" (i.e. one that is not subject to further explanation). He wrote then:

> I do not wish to lay much stress on the greater simplicity of the view of a few forms, or of only one form, having been originally created, instead of innumerable miraculous creations having been necessary at innumerable periods; though this more simple view accords well with Maupertuis' philosophical axiom of "least action."

Does the convergence of Darwin's and Maupertuis' argumentation (cf. the beginning of this subsection) result only from the similarity of the matter at hand, or did Darwin actually read Maupertuis' work?

---

[84] Maupertuis, *Accord de différents loix de la nature*, retrieved at http://fr.wikisource.org/wiki/Accord_de_différentes_loix_de_la_nature_qui_avoient_jusqu'ici_paru_incompatibles.
[85] John Murray, London 1868, p. 12.

Chapter 5

The taming of chance

1. Why does chance need taming?

The Taming of Chance is the title of a well-known book by Ian Hacking (repeatedly quoted in previous chapters). The English word *taming* carries strong associations of 'getting something under control' or 'training or breaking in.' It is not enough to get used to chance, to accept that it exists. It needs taming much as a wild beast that must be forced to obey.

Random events interfere in our lives and may even ruin them. At times, they bring something good, but for the most part we tend to be afraid of random disasters rather than expect a stroke of good luck. Random events cannot be helped since they are unpredictable. Someone takes a walk in the city centre thinking about a holiday beginning next week and suddenly a brick falls on his head. Can one ever get used to such events?

Random events not only interfere in our lives but also act upon the world that surrounds us. Under such circumstances, we may look at them maybe from a little more distance. A stone kicked by accident triggers an avalanche, an accidental gust of wind blows over a tree. Even when random events do not concern us directly, we tend to associate them with risk, with something that can be dangerous.

Quite often, the task of taming something dangerous falls to philosophy. One may venture a statement that philosophical reflection came into existence in order to break in, to tame the dangerous monster known as life: As a result, it makes sense to make the effort to tame chance in the same way.

Aristotle, the Philosopher with a capital P, gave up at the very beginning. In his opinion, "the accidental, then, is what occurs, but *not* always *nor* of *necessity, nor for the most part (...)*" This marks the transfer of the popular understanding of chance onto the ground of philosophical reflection. But science (in Aristotle's times virtually identified with philosophy) deals with what happens regularly and by necessity, so science cannot deal with chance. Science investigates causality, while chance, interfering with the natural order of events, destroys causality, and therefore cannot constitute an object of rational analysis. Our instinctive fear of random events that threaten our daily routine gained philosophical sanction.

Life, however, forces us to face up to unexpected events. A doctor at the bedside of a sick patient and a judge during circumstantial proceedings face problems in which chance and incomplete knowledge appear to reign supreme. And even though in every individual situation the possibility of making an error is considerable, long sequences of analogous situations doubtless exhibit certain regularities. Practice outstripped the philosophical reflection on that score: ancient people developed a number of effective rules that worked well in medicine and in meting out justice.

2. Random events versus the strategy of creation

When Christian thought stepped onto the stage of the ancient world, it introduced critical changes to the issue of chance. The world ceased to be governed by blind fate and became the work of God completely dependent on His will. At the same time, the Greek idea of rationality was not abolished; on the contrary, it was strengthened by becoming ingrained in the rationality of the Creator. Random events, understood as breaches in rationality, can exist only from the human vantage point by virtue of his imperfect understanding (or rather boundless ignorance) of God's intentions. But the problem remained both with respect to the human perspective and with respect to God.

Man is almost incessantly beset by ethical problems: how to proceed in a specific situation when it is impossible to predict the consequences of ones actions in an unambiguous manner? Reflection over this question would soon introduce new methods into the analysis of sentences concerning random situations.

As regards God, the issue of chance is transformed into the issue of contingency. God created the world, but did He have to create it as it is or could he have created it otherwise? This purely theological problem has its consequences for the scientific investigation of the world. If God could not have created a different world, then He is, in fact, just another name for Greek 'necessity' and the logical penetration of necessity constitutes the most fruitful method of studying the world. On the other hand, if God was only driven by His own will in the choice of one world over another, such penetration of logical necessities leads nowhere; it is necessary simply to watch in, all possible ways, how the world operates and draw conclusions from the observations. That is what the empirical method consists in. As is known from history, this method proved to be remarkably effective and we owe all the modern sciences to it.

And what about chance? Can the experimental method be applied to the study of chance?

3. Experimenting with chance

It appears that there are certain situations in which we can experiment with chance. The best opportunity arises when a persistent gambler wants to prevail over chance. Gerolamo Cardano and Chevalier de Méré were such individuals. When the latter was unable to cope with the tricks of chance, he summoned Pascal for help. Pascal, in turn, apparently not quite trusting his own intellect, wrote to Fermat. The taming of chance entered another stage.

In the works of Huygens and later Jacob Bernoulli, it became clearer and clearer that the only method respected by chance was the mathematical method of research. Successes achieved along this route constituted, in a way, an experimental proof against the Aristotelian thesis that chance broke loose from the rigours of rationality. The fact that Aristotle formulated such a statement did not reflect the power of chance, but the weakness of rational methods adopted by him. He thought that sciences concerned with the world might only partly make use of mathematics (e.g. astronomy, optics, acoustics), because mathematics was too weak to be able to capture all the wealth and complication of the physical world. In particular, random events happened "neither always nor by necessity," but simply "happened" and therefore they could not be captured by means of any mathematical formulae. Here is where experimentation with chance helped. It was enough to observe certain types of random events, e.g. results of dice rolls, for a sufficiently long time, in order to ascertain that in 'the distribution of chance' there occur certain simple regularities, e.g. as the number of dice rolls increases, the relative frequency of results obtained approaches a certain constant. Moreover, expressing this fact in the language of mathematics did not demand any extremely advanced methods at all. The main difficulty lay not in mathematical resources, but in the skill of making observations and extracting from them what was important from among the chaos of insignificant detail.

A good example was Bernoulli's limit theorem, the first important result obtained in the mathematical probability theory and its application to the taming of chance. This very distinction is very important – mathematical theory and its uses. In Bernoulli's *Ars conjectandi*, this distinction was not yet clear: Bernoulli simply studied random events that occurred in the real world. Nonetheless, his theorem was a mathematical one: if certain purely formal conditions were met, then the conclusion was correct. This was the case completely irrespective of what happens in the physical world. The theorem itself had nothing to do with the uncertainty of expectations usually associated with gambling. It was based purely on mathematical deduction. But this theorem could be applied e.g. to games of chance or to the mortality rates during the plague in London. It was the case because these random events respected certain regularities that met the conditions of appropriate theorems (e.g. Bernoulli's theorem).

The fact that results of games of chance and mortality statistics during an outbreak of the plague meet the conditions of certain theorems does not follow from mathematical deduction, but constitutes a certain property of the world.[86] The world turns out to be mathematical not only when it is being studied by means of determinist laws of Newtonian mechanics, but also in its own random and probabilistic behaviours.

4. Morality versus probability calculus

Reasonings based on probability not only concern problems related to the physical world, but also constitute an important part of our internal world. As a matter of fact, at almost every turn we must make – often semiconsciously – decisions based on incomplete information or only on more or less vague clues. This also applies to many important decisions in life which we undertake without a full awareness of their present and future circumstances. If rationality even in such situations is to be preserved, certain rules of conduct based on probability should be worked out.

At this point, two polar opposites that affect all kinds of interpretation of probability come to the fore. Probability may be understood objectively when it refers to events that occur independently of our will, or subjectively, when we try to draw inferences from uncertain (yet probable) premises. These are only two extremes with

---

[86] Obviously, no properties of the world (such as the results of games of chance, mortality rates etc.) can meet the conditions of theorems, but the distinction between the properties of the world and their formal model (which can meet the conditions of theorems) was to be introduced much later.

an entire spectrum of approaches in between. For example, the decision whether or not to interrupt a game of chance comprises a mixture of objective probability that can be assessed on the basis of the earlier course of the game with subjective probability associated with the decision whether or not it makes sense to undertake further risk.

Port-Royal Logic grew out of the human longing for a rational order for our beliefs. No wonder that in Port Royal circles primacy was given to religious and ethical beliefs. The greatest gamble of all was human life with itself at stake. Could we not apply here the same rules that had worked for ordinary games of chance? Pascal's wager was not only an exercise in inductive logic or an innovative harbinger of the game theory, but also reflected Pascal's existential struggles with himself. In its written form, it looked like an impartial weighing of arguments, but this impartiality was also a result of a strategy of commitment.

Yet it is not only in the important problems in our lives which demand that we think 'in a limit, when one variable approaches infinity.' When faced with numerous ethical dilemmas we make decisions based on probabilistic arguments. From the historical point of view, it is interesting that it was the theological and ethical speculations of the Spanish casuists that inspired certain considerations connected with probability calculus.

5. Theology – science – theology

As we have already noted on numerous occasions in previous chapters, in those days theological considerations often exerted a strong influence on the positing, and sometimes on the solution, of various purely scientific problems. By the same token, the development of science kept creating new theological problems. Bernoulli's proper understanding of probability was influenced not only by considerations in the area of moral theology, but also by his Calvinist beliefs in rigorous determinism and predestination, which created a conducive conceptual milieu for the formulation of his law of large numbers. But even Leibniz had serious theological difficulties with the reconciliation of even a probabilistic prediction of the future with the contingency of the world. Naturally, all this had far-reaching consequences for the conception of God itself. Does God know the future on the strength of rigorous determinism, i.e. knowing the present state, can He simply compute all that will happen later? Or else, does He know the future by inspection: existing outside time, does He contemplate the entire spectrum of time from the minus to the plus side of the temporal infinity?

A characteristic feature of science is that quite often, it might seem, issues that appear remote with respect to each other begin to interact with one another and generate new 'problem situations.' Apparently, Maupertuis' principle of least action had no connection with philosophical problems of probabilistics, yet he did notice such a relationship or even outright 'provoked' it. Probability calculus had already been developed well enough to warrant the perception of the entirety of natural phenomena in its light. At that time, the teleological argument for the existence of God was very influential. It was acknowledged by Newton and by many scholars from Royal Society circles. Supporters invoked the fact that new empirical sciences showed the traces of purposefulness more and more fully in the world. This view gave rise to an entirely distinct way of thinking called physico-theology. Maupertuis understood that instead of trying to discern everywhere the traces of a purposeful creation, it would be enough to adopt the operation in nature of purely chance processes subject only to probabilistic regularities and the operation of what later Darwin would term adaptation and natural selection.

Still, Maupertuis was a child of his own time and could not free himself from thinking in physico-theological terms. A solution was suggested to him by his principle of least action. The Highest Intelligence, instead of directly interfering in the emergence of each new being or species, should rather establish a single law to solve the matter at a single stroke. Such a law was, according to Maupertuis, constituted by the principle of least action.

6. Perspectives

We can see now how long was the route covered by philosophical reflection since Aristotle. Chance ceased to be a breach of rationality – it became not only a creative element of the order of the world, but also found its own place in God's Great Design. In this respect, the views of today's proponents of the so-called Intelligent Design constitute a serious throwback. In trying to minimise the role of chance in the contemporary theory of evolution in order to underscore part of Intelligent Design (a return to physico-theology), they invoke (maybe even unconsciously) Aristotle's idea that chance cannot be reconciled with the Intelligent Design. This does not mean that I advocate the return to Maupertuis' strategy or a kind of its modification. Maupertuis was only a stage in a story that is still unfolding.

It is not my intention to compile a history of probability calculus. In previous chapters, I have cast only a very selective glance at the first stages of this story. I did this only in order to formulate the issue that I called at the outset 'philosophy of chance'. The problem has been posed. Certainly, one can delve in history decade after decade to look at subsequent intricacies in the development of philosophical notions related to randomness and probability, but this is not, I repeat, my intention. Once the issue has been defined, one should work towards its solution. Indeed, new things started to happen when, in the first half of the 20th century, probability calculus gained enough maturity to become a rightful branch of modern mathematics. It was, as if, a crowning point of the entire previous history of mathematical reflection on probability, and, at the same time, the point of origin of new philosophical problems, yet by no means the final stage in the evolution of probabilistic notions. On the contrary, it is only now that the perspectives of further amazing generalisations have opened up. This process is underway and urgently demands philosophical reflection.

PART TWO

THE FUGUE

The prelude is already behind us and it is time to move on to the key part of the book. Often, the prelude constitutes an introduction to a chamber sonata or a suite, but what I have introduced in the preceding chapters has all the makings of a fugue, an intricate musical form with a clearly presented subject, a unifying motif for a variety of elements.

A fugue is governed by its own internal logic which, however, does not prevent creativity. This was the case with probabilistic reflection in the early 20th century (Chapter 6). Probability calculus had gained enough maturity not only to be effectively applied to straightforward statistical analyses such as the mortality rate or annuities, but also to subtle problems in astronomy and statistical physics. Practical applications often show a new face to theoretical notions and mathematicians noticed that the notion of probability was related to the mathematical concept of measure. This observation proved to be critical and bore fruit in the form of the formalisation of probability theory by Kolmogorov. The classical probability theory achieved mathematical maturity and immediately began to produce results in the form of new important insights in mathematics as a whole. Interpretative matters remained. Frequency, Bayesian and other interpretations have had a relatively minor impact on the development of mathematics, but they continue to play an important role in philosophical considerations.

Interpretations of probability can be divided into two groups: those that perceive the sources of probability in the subjective absence of necessary knowledge, and those that discern such sources in the objective properties of the world (Chapter 7). We are especially interested in the latter. From this point of view, the analyses of chance conducted by Marian Smoluchowski are a case in point. The laws of physics express themselves through differential equations, while the solutions to these equations determine the possible behaviours of physical systems. A specific solution is selected by the initial conditions (or boundary conditions), but they constitute a random element with respect to a given law of physics, yet without this random element the laws of physics would not operate. Within the network of laws of physics, there are places open to the operation of chance and the number of these places is exactly what it should be.

Why is probability calculus so effective in the study of laws of physics and their practical application to the world? Surely, it would not be the case if the world did not have the property of frequency stability. Thanks to this property, in long series of tosses of a fair coin we obtain more or less the same number of heads and tails. This is a property of the world, not of probability calculus. The applicability of probability calculus to the world poses a serious philosophical problem.

The formalisation of a mathematical theory does not have to entail its closure since practical applications may push for its further development. Probability calculus found its own new domain of application in quantum mechanics (Chapter 8). Almost from the very beginning, it was known that it was a probabilistic theory, but soon it transpired that the principles of probability calculus operate in it in a slightly different way from their classical applications. Thanks mainly to the works by von Neumann, it became clear that probabilistics encoded in the mathematical structure of quantum mechanics was, in fact, a new quantum probability theory. Several decades had to elapse before we realised that this fact opened up the way to further generalisations of the mathematical measure theory. This chapter in its mathematical dimension is doubtless more difficult than the other chapters. In principle, it can be left out without substantial prejudice to the understanding of the main train of thought in the book, still I encourage the Reader to try and at least intuitively grasp its basic ideas. The development of science reaches out for ever more and more sophisticated exploratory tools and conceivably the next stage in the history of issues discussed in this book will develop in the direction heralded by this chapter.

If a random event is to be understood as one whose (a priori) probability is less than one (because we would not be inclined to call an event random whose probability equals one), and we have a number of kinds of probability calculi, which one should we use to deal with chance (Chapter 9)? As regards chance in the macroscopic world, undoubtedly the classical probability calculus is the right one, but what is essential for the Universe is the fundamental level. Is it probabilistic? If so, in the sense of which probabilistic measure theory?

Chapter 6

The probabilistic revolution

1. Introduction

The opening decades of the 20th century in Europe saw several crucial scientific revolutions. The most important include: the invention of the theory of relativity and quantum mechanics as well as changes that took place in the foundations of mathematics crowned by Gödel's theorems. Somewhat less spectacular, but – as was to be soon shown – not less far-reaching was a revolution affecting the very foundations of probability calculus.

Probabilistics played an important role in classical physics, e.g. in the discussions of errors in measurement, and, together with the emergence of thermodynamics and statistics, physics became one of its basic tools. The invention of quantum mechanics changed the status of the notion of probability. While preserving its instrumental function, it became one of the basic explanatory categories in physics. It was the case not only in quantum physics: the invention of non-linear thermodynamics and the theory of non-linear dynamical systems showed a completely new role to be played by probability theory in the domain of classical physics.

In this chapter, we offer a fairly concise review of changes in thinking about probability in the 20th century, which combined to form what we called the probabilistic revolution. Indeed, this review is intended as a historical one, but its main objective is not so much to discuss history itself[87] as to prepare the ground for a deeper understanding of certain philosophical aspects of probability-related issues. Our historical review in principle finished with Kolmogorov's work, which provided a solid mathematical grounding for probability theory. Then, we will only sketch a handful of several philosophical comments and remarks pointing to the future.

In the title of this chapter, I used the word 'revolution.' In what sense can what occurred in probabilistics in the 20th century be called a revolution? A similar question was asked by Ian Hacking, but with respect to an earlier period, when the concept of probability was emerging from various practical endeavours and had only just started to assume a more mathematical shape.[88] Hacking notes that this was not a revolution in the Kuhnian sense, where periods of normal science are interspersed with changes in paradigms. He also quotes I. B. Cohen, who believed that in the case of probability, it had been 'a revolution' in application rather than a wholesale scientific revolution. Hacking does not subscribe to such an opinion. In his view, it was not only an issue of application, because, at the time, such key concepts were being shaped as chance, determinism and ideas connected with probability that took root in our present way of understanding of the world. It was thus, as Hacking's believes, a revolution in the colloquial sense of the word, a revolution more fundamental than those written about by philosophers of science. I think, though, that Kolmogorov's achievement and all that 'surrounded' it can be called a revolution in Kuhn's sense. One may also venture a statement that it was a revolution in a peculiar sense, since, although be-

---

[87] An extensive treatment of the subject of history of probability appears in J. von Plato, *Creating Modern Probability. Its Mathematics, Physics and Philosophy in Historical Perspective*, Cambridge University Press, Cambridge 1994. To a considerable extent, the present chapter draws on this book. References missing from it should be sought in von Plato's work. Another useful reference is M.C. Galavotti, *Philosophical Introduction to Probability*, CSLI Publications, Stanford 2005.

[88] I. Hacking, *The Emergence...*, *op. cit.*, Introduction.

fore Kolmogorov there existed a number (at least several) paradigms, his work resulted in the domination of a single paradigm. It is worth looking at these changes in a little more detail.

2. Probability and astronomy

Before the beginning of the 20th century, the situation in probabilistics was to some extent paradoxical, since it can be maintained that at the time, no mathematical probability theory yet existed, unlike its practical applications. As we already know, probability calculus was born out of a reflection on games of chance and preserved something of such provisional immediacy. Its basic notions and methods were developed in connection with specific problems; with references to other branches of mathematics being few and far between. An important area of the application of probability calculus was in statistical physics and this is where the most significant theoretical advances were made. This proved to be true to such an extent that probabilistic methods appeared to be more of a tool in the hand of physics than elements of a mathematical theory.

Around the turn of the century, changes in this area began to accumulate: gradually, relationships came to light amongst probability calculus and certain problems in pure mathematics. An essential element in these changes was the development by Borel and Lebesgue of the mathematical measure theory. On the one hand, the theory made it possible to express certain probabilistic notions in the language of measures applied to different sets, while on the other hand, it revealed probabilistic significance of some well-known mathematical structures.

One of the first hints of this kind came from a completely unexpected source. The Swedish astronomer Hugo Gyldén (1841–1896) investigated the motion of planets around the Sun. These motions are described by quasiperiodic functions. Gyldén used a method de-

veloped by Lagrange of approximating these functions using continued fractions. As we know, every real number can be represented as
a continued fraction:

$$a_0 + \cfrac{1}{a_1 + \cfrac{1}{a_2 + \cdots}},$$

where a_i, $i = 0, 1, 2,...$ are integers. Gyldén noticed, even though he
had no proof, that in the distribution of numbers $(a_0, a_1, a_2,...)$ certain regularities appear if the number corresponding to the continued fraction is irrational. Using this as his starting point, he tried to
show that probability of a certain sequence representing disturbances in the motion of a planet being divergent is smaller than any chosen value. Gyldén did not succeed in proving it, but the idea that there
may exist 'negligibly small' probabilities turned out to be important.
This insight was clarified by Poincaré in his work on the motion of
three bodies, where he formulated a theorem later called Poincaré recurrence theorem. It says that in a compact phase space, every trajectory returns arbitrarily close to a state that it already passed and that
exceptions to this rule are possible, but 'negligible.' Poincaré wrote
that such exceptions have a 'zero probability.' Later, after the development of the measure theory, such exceptions would be assigned the
measure zero.

Gyldén's work passed unnoticed until 1900, when two Swedish
mathematicians, Torsten Brodén and Anders Wiman, started discussing it. Brodén was probably the first mathematician to have perceived
a direct relationship between probability theory and Borel's measure
theory, but Wiman proved to be the first one to have applied measure
theory in practice to determine the probabilities suggested in Gyldén's analyses. The works of these two mathematicians, and thanks
to them, also Gyldén's work, have made a significant contribution to
the development of probability theory.[89]

---

[89] More information on the subject can be found in J. von Plato, *op. cit.*, pp. 6–8, 27–32.

3. Towards measure theory

Admittedly, the famous list of unresolved problems in mathematics presented by Hilbert in his address to the International Congress of Mathematicians in Paris in 1900 had a huge impact on the later development of mathematics. In his sixth problem, he postulated "to treat in the same manner, by means of axioms, those physical sciences in which already today mathematics plays an important part." Under this heading, Hilbert included probability calculus and statistical mechanics. Soon several papers were published (authored by, among others, Rudolph Laemmel and Ugo Broggi) that took up Hilbert's challenge, yet they did not effect the expected breakthrough.

When Hilbert put forward his postulate, a tool had already been in existence that could be used in order to endow probabilistics with the shape of a mathematical theory in the full sense of the word. There had already existed attempts to use the tool accordingly. The mathematical measure theory – because this is what we are talking about – began to take shape towards the end of the 19th century, mainly for the needs of analysis and integration theory. In 1888, Émile Borel defined measure as a generalised notion of the length of a line segment. The measure of an interval $[a,b]$, $a < b$, on the real line $\mathbf{R}$ is $b - a$. The measure of a set that consists of a sum of a countable (or finite) number of such intervals is the sum of measures of these intervals. If Y is a subset of set X, then the measure of the difference $X - Y$ in X is the difference of their measures. Soon afterwards, H. Lebesgue (1904) generalised Borel's definition and applied it to his integration theory, and in 1915 M. Fréchet gave measure theory its present abstract shape.

In his original work of 1905 on measure theory, Borel suggested that it could be applied to analysing probabilistic problems in classical mechanics, but the first serious step in this direction was not made until 1913 by M. Planckerel. A year later, Felix Hausdorff in his work *Grundzüge der Mengenlehre*, which soon became a standard manual of the set theory, noted that the normalised measure from a formal point of view has all the features of probability (though he did not yet give a formal definition of probability as a measure). This was followed by numerous publications whose authors applied the concept of measure to problems connected with probability, but complete success came only two decades later.

4. Statistical mechanics

Statistical mechanics was born around the mid-19th century out of considerations on the theory of heat. This theory soon adopted two forms: the phenomenological (macroscopic), based on directly observed quantities, and the statistical, which focussed on the motion of particles subject to the laws of mechanics. From the very beginning, the junction of these two notions was beset by serious problems. In the phenomenological approach, thermal processes demonstrate a distinct directionality (with respect to time), while the statistical approach is exactly symmetrical in time. In order to solve these problems or at least understand them, scientists invoked arguments related to probability, since from the vantage point of the statistical approach, observable (and hence macroscopic) quantities cannot be anything other than the results of certain averaging of statistical processes.

The first application of probability tools to the theory of heat appeared in the works of Augustus Krönig (1856) and Rudolph Clausius (1857). It was, in a way, the first approximation, but Clausius soon refined his methods. It is interesting to note the affinity between the use of probabilistic calculations in astronomy and in theory of gases. Nineteenth-century astronomers assumed that at a given moment a celestial body under examination occupied a certain "true position" and any inaccuracies in its determination were due to observational errors. In this spirit, they applied probability calculus to the discussion of errors. Adolph Quetelet and John Herschel were the first to take under consideration the actual disturbances of position, not only observational errors. This point of view was adopted by Maxwell, when in 1860 he derived his famous particle distribution function. He treated thermal motions of particles as a physical process subject to laws of statistics. Maxwell's views to a certain extent evolved towards a particular kind of indeterminism. At first, he thought that gas particles moved in compliance with the laws of mechanics, except at the moment of collision, when these determinist laws of mechanics were replaced by probability, but already in 1875 he wrote that thermal motions were so irregular that "the direction and magnitude of the velocity of a molecule at a given time cannot be expressed as depending on the present position of the molecule and the time."[90] Classical mechanics thus required a supplementation that to a significant extent (not only as a result of our ignorance) invoked the laws of statistics and probability.

Such views in the domain of statistical mechanics caused Maxwell to entertain more and more clearly an indeterminist view of physics as a whole. He maintained that the principle that "the same causes will always produce the same results" constitutes a metaphysical doctrine that is not corroborated by science, since instabilities of certain processes may cause discontinuities of time. Such a discontinuity occurs when small perturbations in causes produce large perturbations in results. These instances of discontinuity were called 'singularities' by Maxwell who thought that during more complex phenomena such singularities may appear 'very densely,' whereby the analysed system becomes unpredictable. For that reason, Maxwell maintained that once physics has penetrated the 'invisible motions' of the microworld, it may turn out that his magnified image will not be the same as the world of Newtonian physics. His comment also pertains to the development of physics as a science: its future state will not be only an 'enlarged picture' of its state in the past.

---

[90] J.C. Maxwell, *On the Dynamical Evidence of the Molecular Constitution of Bodies*, "The Scientific Papers of James Clerk Maxwell," vol. 2, University Press, Cambridge 1890, pp. 418-438, quotation from p. 436.

Another key figure in the development of statistical mechanics was Ludwig Bolzmann. He characterised the difference between his own and Maxwell's approaches to statistical issues, noting that while he himself determined the probability of a certain state of a given physical system invoking the average period of time during which such a system remains in this state, Maxwell preferred to analyse an infinite set of systems with all possible initial conditions. Bolzmann's approach led to the formulation of the ergodic theorem.[91] Maxwell's approach was associated with the concept of the so-called *ensemble*. The concept is born in the following situation. We are examining a single specific mechanical system, but we do not exactly know its state and thus we are considering a set – *an ensemble* – of all possible systems identical with the one under investigation in respect of its properties as we know them. The extent of our ignorance concerning the system under investigation can be expressed with the aid of the measure on the *ensemble*.

The *ensemble* method played an important role in Gibbs's research. It accorded with his philosophical preferences, since Gibbs supported the epistemological interpretation of probability, i.e. the view that the necessity to use probability stems from our ignorance of the system under consideration. In his opinion, probability describes our ignorance "as something taken at random from a great number of things which are completely described."[92] It is in the spirit of this philosophy that the *ensemble* method can be interpreted.

Bolzmann produced a consistently statistical interpretation of the second law of thermodynamics. At first, it attracted numerous objections. The problem was – as I mentioned above – how to rec-

---

[91] It states that a single trajectory of a mechanical system fills up the entire phase space, which causes that time averages measured along the trajectory equal the averages on the phase space (up to zero measure sets). For exact definitions of concepts related to ergodicity cf. e.g.: W. Fomin, I.P. Kornfeld, J.G. Sinaj, *Ergodic Theory*, Springer-Verlag, New York 1982, Chapter 1.

[92] J.W. Gibbs, *Elementary Principles in Statistical Mechanics*, Dover, New York 1962 (first edition 1902), p. 17.

oncile the irreversibility of thermal processes at the macroscopic level with the reversibility of statistical phenomena at the microscopic level. The dispute between Bolzmann and Zermelo is widely known. The latter contrasted macroscopic irreversibility with Poincaré recurrence theorem that states – as we already know – that the trajectory of a mechanical system will, after a sufficiently long time, return arbitrarily closely to its initial state.[93] The charge was incomprehensible insomuch as Poincaré himself, while introducing his theorem, indicated that even though the theorem applies with probability one, exceptions cannot be excluded, even though one should ascribe to them a zero probability. Today, we would say that the set of exceptions is a zero measure set. Bolzmann in his reply even more emphatically explained the statistical character of laws of thermodynamics. Everything points to the fact that Zermelo finally accepted Bolzmann's arguments. Although he did not admit it publicly, in his postdoctoral thesis of 1899, he justified the probabilistic foundations of dynamics.

Bolzmann pointed out that for a sufficiently small number of gas particles, the statistical character of the laws of motion should become apparent in the fluctuations around the mean value. This observation also had its repercussions in the discussion concerning the thermal death of the Universe. Following the suggestion of his assistant, Schuetz, Boltzmann put forward a hypothesis that if we assumed that the Universe was sufficiently large, then there was a finite probability that, although the Universe as a whole finds itself in a state of thermodynamic equilibrium, a part observed by us constitutes a local fluctuation where thermodynamic processes develop toward a greater entropy.[94]

---

[93] I write about this issue more extensively in *Idea wiecznych powrotów: od Zawirskiego do dziś (The Idea of Eternal Recurrence: from Zawirski untill Today, in Polish)*, *Krakowska filozofia przyrody w okresie międzywojnnym, (The Krakow Philosophy of Nature Between the World Wars)*, vol. II, M. Heller, J. Mączka, P. Polak, M. Szczerbińska-Polak, eds., Biblios – OBI, Tarnów – Kraków 2007, pp. 281–303.

[94] Cf. my article *Zagadnienia kosmologiczne przed Einsteinem (Cosmological Issues Before Einstein, in Polish)*, "Zagadnienia Filozoficzne w Nauce" 2005, 37, pp. 32–40.

Antientropic processes, though statistically possible, are very unlikely. In his dispute with Zermelo, Bolzmann estimated that for one cu-
bic centimetre of gas the length of time needed for it to return to its initial state (in compliance with Poincaré's recurrence theorem) is ca. 10^{19}
seconds. Are, then, the fluctuation phenomena discernible at all? Soon it
transpired that, in fact, they had been observed for a long time. Now they
became a rewarding area for the applications of probability calculus.

5. Brownian motion

Brownian motion was observed for the first time in 1827. Some time
later it was interpreted, purely qualitatively, in the spirit of the kinetic
theory of gases, but a complete theoretical account of the process was
not provided until 1905 (by Albert Einstein) and 1906 (by Marian Smoluchowski). Smoluchowski's approach was based on the analysis of the
collisions of suspension particles with fluid particles whereas Einstein
saw this phenomenon as a random process occurring in continuous time.
Even though Bolzmann had already recognised that such an approach
was possible, but it was Einstein who presented it in the form of a mathematical model. This was a pioneering approach, because it was not until the late 1920s that mathematicians took an interest in this problem,
and the theory of random processes in continuous time did not acquire
momentum until the publication of Kolmogorov's seminal work. From
the vantage point of these later developments, Smoluchowski's approach
to Brownian motion could be viewed as Markov's processes in discrete
time.

Random processes in continuous time entailed certain philosophical
difficulties. At the time, the frequency interpretation of probability was
universally accepted, so it was only natural to imagine a random process
as a series of random events operating in discrete (although maybe extremely minute) segments of time. Chance operating for some time in a

continuous manner was difficult to grasp intuitively. Besides, processes taking place in continuous time were traditionally considered to belong to the domain of mechanics, which was deterministic, not probabilistic.

Einstein's work on Brownian motion doubtless weighed heavily on his later views on the role played by probability in physics and on the issue of interpretation of quantum mechanics. In his papers, there are distinct traces that his own views on probabilistics were based on his reading of Bolzmann's works, but the knowledge thus gained he later 'filtered through' (so to say) specific calculations, which had to be performed in order to explain Brownian motion, and it was those calculations that finally contributed to the crystallisation of his views. It is also worth mentioning that Einstein's work on Brownian motion not only swiftly won recognition, but also contributed to the development of molecular statistics, which, in effect, helped to overcome the last vestiges of physicists' resistance to the hypothesis of atomism.

Einstein's famous saying that God does not play dice, should not be understood as a protest against the use of probabilistic methods in physics. What Einstein wanted to say was only that the role of probability in quantum mechanics should be analogous to its role in classical statistical mechanics, i.e. the present quantum mechanics should be related to the future fundamental theory more or less in the same way as statistical mechanics was related to classical mechanics. Einstein found it hard to accept that single physical units (elementary particles) might behave probabilistically, but, in principle, he had nothing against the application of probabilistic methods to the *ensembles* of physical units.

6. Kolmogorov

A turning point in the development of probability theory was the publication in 1933 by Andrey Kolmogorov of a book titled *Grundbegriffe der Wahrscheinlichkeitsrechnung*.[95] This work outdistanced

---

[95] Springer, Berlin.

numerous previous attempts to define the notion of probability and axiomatise its theory. It also became a reference point for practically all of the later major works in the more broadly conceived area of probabilistics. Kolmogorov had already published several works in the domain of probability and in some of them he made use of measure theory. His other works were in logic, functional analysis, classical dynamics and turbulent flow. He was also keenly interested in the foundations of mathematics. He believed in intuitionism and constructivist methods and these preferences were clearly revealed in his approach to probability calculus.

The issue of interpretation of probability was considered by Kolmogorov as the relation of the mathematical notion of probability to phenomena and processes that occurred in the physical world. He was inclined to adopt the frequency interpretation of probability, and in the 1960s made a very decisive step in this direction. He was also interested in the question whether the random variation of an event implies that this event is not subjected to any laws at the deepest level, and he was inclined to accept the opinion that probability could be assigned an objective meaning without resolving the issue of randomness versus non-randomness at the fundamental level, since the statistical behaviour of a given event may constitute an end result of a collection of specific circumstances on which it depends.

Usually, mathematical structures relate to the world of empiricism via theories or physical models. Probability calculus appears to be an exception, since its principles are applied directly to physical processes, e.g. when rolling a die or flipping a coin. With his distinctive accuracy, Kolmogorov noted that in mathematical probability theory we never deal directly with a physical reality, but only with its models (patterns) that we build ourselves. One may say that the mathematician, when constructing such a model, in a way takes over the role of a physicist.

Classical mechanics contemplates deterministic processes, i.e. those in which a present state of a given system determines its behaviour in the future, but it may equally well be the case that that the state of the system at the present moment determines only the probability distribution of its possible states in the future. Such processes are called stochastic. In such processes what is determined are not the subsequent states, but subsequent probabilities of occurrence of those states.

Kolmogorov took up Hilbert's challenge (the sixth problem from the list of unsolved problems in mathematics) to axiomatise probability calculus. A formalisation of probability theory captures only the formal relationships within a logical structure, abstracting from concrete situations, consequently, the price to be paid for a good formalisation is that a ready formal system, by becoming part of pure mathematics, admits of more than one interpretation (Skolem theorem). Probability calculus understood intuitively applies only to random phenomena. A formalized probability calculus applies to formally defined 'events' that may have nothing in common with randomness. As an example of such a situation Kolmogorov quoted the study of the sequences of digits in decimal expansions of different numbers. Such distributions, being established in advance (after all, they are computed with a rigorous accuracy), do not result from the operation of chance, yet the axioms of probability calculus apply to them.

Unquestionably, Kolmogorov's fundamental achievement was to axiomatise probability theory, yet his work also brought about two other important developments: the theory of stochastic processes and the theory of conditional probabilities. Hence Kolmogorov's work represents not only a crowning achievement and a supplement to earlier research (which, as if in its shadow, lost its relevance), but also marks the beginning of a chain of new research for which it set requisite standards. Since Kolmogorov's time, it has become hard to imagine a mathematically responsible concept of probability without reference to measure theory. To be sure, it needs to be understood properly: in research, the approach associated with measure theory was applied almost immediately after the publication of Kolmogorov's work, but it was not permanently included in textbooks until after World War 2.

7. Interpretations – von Mises

Before Kolmogorov, the frequency interpretation of probability dominated the scene. This common-sensical approach was elevated to the status of a philosophical interpretation by Richard von Mises. Professionally, he dealt with the applications of mathematics and his interpretation of probability calculus bears distinct features of such a practical approach. Philosophically, he belonged to the neopositivist movement. One of his books devoted to probability calculus – *Warscheinlichkeit, Statistik und Wahrheit*[96] – was published as the third volume of the series *Schriften zur Wissenschaftlichen Weltaufassung* issued by the Vienna Circle.

In keeping with positivistic philosophy, von Mises tried to capture the idea of probability by way of abstraction and idealisation of well-determined observable phenomena. He applied the method of informal axiomatisation, calling his axioms 'conditions.' In his opinion, probabilities should not be 'measured' by referring to certain abstract possibilities (such as 'other worlds'). Every event and every sequence of events that could be parameterised by means of labels constituted a certain mathematical space and were 'vehicles' for probabilities.

Probability theory in von Mises' interpretation referred only to events repeated a large number of times. If this condition was not met, then the problem was considered to be badly expressed in probabilistic terms. Likewise, the issues connected with subjective probability were considered by von Mises as unfit for quantitative expression, and thus scientifically unapproachable.

In von Mises' approach, the randomness of events is postulated by theory. However, the laws of classical mechanics do not involve the element of randomness, and consequently, it may not be considered an area of operation of probability. Yet there also arises the question strongly emphasised by von Mises, namely whether the laws of classical mechanics could actually explain all kinds of motion that occur in nature.

---

[96] Springer, Vienna 1928.

Authentic probabilistics demands phenomena for which no 'algorithms' exist. One may surmise that what von Mises meant by 'algorithms' was simply the absence of computational procedures.

8. Interpretations – de Finetti

By approximately 1930, the subjective interpretation of probability was completely in retreat. The situation both in physics (success of statistical mechanics, development of quantum mechanics) and in philosophy (neopositivism) did not favour subjectivist interpretations. Numerous publications by Bruno de Finetti curbed this trend. In his work, he invoked empiricist and operationalist philosophy. In his view, probability was a subjective degree of someone's conviction about something, something akin to a direct report on a sensory experience. However, the moment such a 'degree of conviction' is expressed, it becomes similar to an officially recorded sentence and can be investigated using 'independent methods.' Its mathematical results, irrespective of their interpretation, boosted confidence in his interpretative views. Independently of de Finetti, similar conclusions were reached around the same time by Franc Ramsey at Cambridge.

The subjective interpretation of probability may be illustrated by the following problem. I am considering a certain hypothesis H, by means of which I wish to explain a certain phenomenon, and I assume that the probability of hypothesis H being true is $P(H)$. How I know it, has no impact on the further course of reasoning; it may be a conclusion based on my assessment of certain arguments in support of hypothesis H or against it, or it may simply be my subjective conviction. De Finetti calls probability $P(H)$ prior probability, but henceforth we will use the term initial probability. Let us assume that result E of a certain experiment has been observed and that there exists a certain probability, let us denote it as $P(E|H)$, that E constitutes a consequence of hypothesis H, in other words, H explains E. Then a question arises, how this new information will influence my belief that hypothesis H should be assigned probability $P(H)$? The new probability that should be assigned to hypothesis H taking E into account, de Finetti calls posterior probability, whereas we will call it final probability. De Finetti further states that, although the initial probability may be based on entirely subjective reasons, the journey from initial to final probability should be subjected to rigorous deduction and should occur according to the following formula:

$$P(H|E) = \frac{P(E|H)}{P(E)} \cdot P(H),$$

where $P(E|H)$ denotes the probability that E is true on the strength of hypothesis H, and $P(E)$ is the probability that E is true without adopting hypothesis H. The first factor on the right side of the above formula is thus the ratio of probability E assuming that H (i.e. the probability that E results from H) to probability of E without assuming that H. If, then, we assume that this factor represents the influence of E on the recognition of H, then the above formula can be rewritten in a shorter form:

final probability = (the impact of E on H) × initial probability

Let us note that all the time we reason here 'against the tide' as is done in empirical sciences: we know the result E of a certain experiment and we seek hypothesis H from which E would result, in other words, one that would explain E.

If new arguments appear in support of hypothesis H, then final probability can be treated as a new initial probability and the entire reasoning can be repeated. [97]

The subjectivist interpretation of probability sometimes is called the Bayesian interpretation. Thomas Bayes lived in the 18th century, was a Presbyterian clergyman and took a deep interest in probability calculus.

---

[97] A simple proof of the Bayesian theorem, which is only a little more complex version of the formula shown above, can be found in W. Janowski, *Elementy rachunku prawdopodobieństwa (Elementary Issues in Probability Calculus, in Polish)*, Biblioteczka Nauczycieli Matematyki, PZWS, Warszawa 1963, pp. 37-41. The book also contains a number of instructive examples.

The natural basis for subjectivist interpretations of probability calculus is offered by classical determinism: determinist laws operate in nature and it is only our ignorance of them that compels us to use probability calculus. De Finetti's view was more sophisticated. Its formulation may have been influenced by the emergence of quantum mechanics, which already at that time strongly undermined the faith in deterministic laws operating at the deepest level of nature. The suspicion that this level was 'inherently indeterministic' gained more and more currency. De Finetti, however, could not agree with this: probability is a measure of the extent of our ignorance, not a feature of the world (at any of its levels). His solution, quite a surprising one, was consistent throughout. The laws of nature, whether those expressed by classical or quantum physics, only have the character of subjective statistical regularities. De Finetti supported his argument with references to positivistic philosophy: from the operationalist point of view, there are no reasons to believe in the objective existence of laws of nature.

9. Probability as measure

We have already said a great deal about probability as measure, now is the time to give a brief overview of some relevant definitions.

Let us consider space X and a set S of its subsets (we assume that the empty set belongs to X). We also assume that set S constitutes the so-called σ-field, i.e. for every two subsets of the set S, their sum and their difference also belongs to S and each countable sum of subsets of the space S also belongs to S. Let there be given the function

$$m : S \rightarrow \mathbf{R}^+,$$

that attributes to every subset s_i a positive real number $m(s_i)$. We also assume that (1) $m(\emptyset) = 0$ and that (2) for each countable family of mutually disjoint subsets $(s_1,..., s_n,...)$ of the space S one has

$$m\left(\bigcup_{i=1}^{\infty} s_i\right) = \sum_{i=1}^{\infty} m(s_i)$$

Function m is called a measure on X, while the value $m(s_i)$ is called a measure of subset s_i. The two above conditions mean that (1) the measure of an empty set equals zero and (2) the measure of the (countable) sum of mutually disjoint subsets of space X equals the (countable) sum of measures of these subsets. The triple (X, S, m) is called measure space.

If the condition that $m(X) = 1$ is met, then measure space (X, S, m) is called probability space (or *probabilistic space*), subsets of space X (i.e. elements of space S) are called events, whereas $m(s_i)$ is the probability of event s_i.

Needless to say, the above constitute only the most important notions developed by Kolmogorov into a full mathematical theory. His axiomatics proved to be so effective that it comprised virtually all previously known formal concepts of probability as its particular cases. As I have already mentioned several times, Kolmogorov's work triggered a rapid development in probability theory, both as a purely mathematical theory and in its numerous applications.

10. Probabilistics versus mathematical structuralism

Thanks to Kolmogorov's work, probability theory took up its rightful place as part of mathematics, moreover, in time it began to play a more and more significant role in it. Contemporary mathematics differentiates between two main types of structures: algebraic and analytic ones. Within the analytic structures, there are two main subclasses: topological structures and structures of measure.[98] Naturally, probabilistic structures belong to the structures of measure. Being part of mathematics understood as a formal science, they are structures in the same sense as all the other mathematical structures.[99] They refer to the physical world via probabilistic models and only then does the problem of their interpretation arise. Can they be interpreted in the spirit of structuralism? Subjectivist interpretations would run up against difficulties in this respect, unless we were to interpret our cognitive categories in a certain (figurative?) structuralist sense, but objectivist interpretations of probability theory perfectly fit in the structuralist pattern of the understanding of physics. We have thus a certain mathematical structure, in this instance a probabilistic one, and we map it on certain features (or aspects) of the world, thanks to which these features (or aspects) of the world acquire structural properties.[100] To be sure, they should not be understood statically (as e.g. structural features in architecture), but dynamically, i.e. in the sense that the property of variability in time is integrated into the structure, and may even play a dominant role in it. Yet this is nothing exceptional in structures that appear in physics. After all, all dynamic theories lend themselves beautifully to a structuralist understanding.

Probabilistic structures in physics are characterised by something else, namely the occurrence of random events. If a random event is to be understood as one that occurs even though it is assigned a small probabilistic measure, then probabilistic structures in physics are characterised by an intrinsic interaction between random and regular events.

---

[98] Cf. e.g. P. Roman, *Some Modern Mathematics for Physicists and Other Outsiders*, vol. I, Pergamon Press, New York – Toronto 1975, pp. XXIV–XXV.

[99] I address the issue of structuralism in the philosophy of mathematics in *Filozofia i Wszechświat (Philosophy and the Universe)*, Universitas, Kraków 2008, Chapter 9.

[100] Cf. *ibid.*, Chapter 10.

11. After Kolmogorov

Even though our review ends with Kolmogorov's work, it does not mean that the probabilistic revolution ended there. What ended was only one of its important stages – the formulation of the fundamentals and the inclusion of probability theory in the mainstream of the development of mathematics. Further history of probabilistics ran along different, but intertwining tracks, including, first of all, the development of probability theory itself as a branch of pure mathematics. This theory, once included in the rapid current of the development of mathematics in the 20th century, acquired its dynamism, and, at the same time, enriched it with numerous creative strands. For example, the interaction between probabilistics and the dynamical systems theory gave rise to the theory of stochastic processes and the theory of ergodicity. Somewhere at the interface of these conceptions the theory of dynamic chaos was born, which proved the indispensability of probabilistic methods in the most traditional divisions of classical mechanics. The above-mentioned theories, despite their names being indicative of their practical applications, have been developing as branches of abstract mathematics and have significantly contributed to the changing face of pure mathematics at the end of the 20th century; the area of its application is less and less reminiscent of an area of 'organised' objects and is increasingly becoming 'a simmering ocean' of dynamic structures.

These changes induced most mathematicians to cease to consider applications of their ideas as a disgrace, on the contrary, today they treat them as a sort of mark of distinction. All the more so that recent history has proved that practical applications often stimulate progress in abstract theories. Not infrequently, such progress consists in revealing completely unexpected directions for further development. It is enough to mention nonlinear thermodynamics and the theory of structure formation (complexification). This impact of interaction between abstract theories and their applications caused that the title of

Prigogine's book *From Being to Becoming*[101] devoted to the issue of self-organization may be successfully related to the above-mentioned transformations occurring in the domain of interest of pure mathematics.

In no book devoted to probability, even if it is supposed to deal only with its classical aspects, can one fail to mention quantum mechanics. At the time when original works by Kolmogorov were published, it was still possible to consider quantum mechanics only as a new domain for the application of probabilistic techniques, soon, however, it became clear that it exploded the classical framework of the notion of probability. Although during an elementary course of quantum mechanics students may still attempt to reason in terms of classical probability calculus, already their first attempts to transfer probability as a measure on operator algebras (which inevitably occurs as a result of a deeper look into quantum structures) entail completely new challenges. No wonder that today quantum probability theory is treated as a generalisation of the classical theory.[102]

It is a well-established mathematical rule that one generalisation entails others. The quantum probability theory makes an extensive use of functional and algebraic methods, no wonder that the next step involved abstracting from situations connected with quantum mechanics and linking the notion of probability with a certain type of algebras, not necessarily commutative ones. Thus, a noncommutative probability theory was born, also known as the free probability theory.[103]

There exists yet another distinct line of research connected with probability theory, namely that of philosophical analyses. It covers such issues as the nature of probability and its different interpretations, probabilistic epistemology and relationships between probabi-

---

[101] Freeman and Co., New York 1980.
[102] Cf. a review article by M. Rédei, S.J. Summers, *Quantum Probability Theory*, "Studies in History and Philosophy of Modern Physics" 2007, 38, pp. 390–417.
[103] Cf. e.g. D.V. Voiculescu, K. Dykema, A. Nica, *Free Random Variables*, American Mathematical Society, Providence 1992; I. Cuculescu, A.G. Oprea, *Noncommutative Probability*, Kluver, Dordrecht – Boston – London 1994.

listics and inductive logic, logical aspects of probability theory and the nature of application of probability calculus to different sciences. Such research invoking classical probability theory already has its own long history and extensive literature.[104] At present, it is conducted more and more often in the area of quantum probability.[105] Noncommutative probability theory adds new questions to traditional philosophical ones, e.g. how to understand probability in strongly nonlocal spaces in which individuals do not exist? This problem area quickly reaches a point where bold philosophical analyses are in order.

---

[104] Cf. M.C. Golavotti, *Philosophical Introduction to Probability*, CSLI Publications, Stanford 2005; D. Gilles, *Philosophical Theories of Probability*, Routledge, London – New York 2008; W. Załuski, *Skłonnościowa interpretacja prawdopodobieństwa (Propensity interpretation of probability, in Polish)*, OBI – Biblos, Kraków – Tarnów 2008. I address certain issues connected with the philosophy of probability in my article *Filozofia przypadku (Philosophy of Chance)*, in "Prace Komisji Filozofii Nauk Przyrodniczych Polskiej Akademii Nauk," Kraków, pp. 57–66.

[105] Cf. the entire issue no 38 of "Studies in History and Philosophy of Modern Physics" 2007.

Chapter 7

What is chance?

1. Introductory remarks

In common parlance, the term 'chance' is ambiguous. It can be made more specific in a variety of ways, but it would be a futile undertaking only for the sake of specificity itself. However, the notion of chance appears in different theories of nature (often in physics and biology), and also in numerous philosophical discussions. And it is in these contexts that the situation begins to be precarious, unless we know what we are talking about, consequently, it is time we devoted more attention to the notion of chance, the more so that we already have sufficient material (outlined in previous chapters) on the basis of which to approach this matter in a responsible manner.

2. Preliminary definitions

There are numerous definitions, explanations and descriptions of chance. One may even risk the statement that all these endeavours attempt to capture or specify the intuition that induces us to talk about chance when we are faced with an event whose probability is (very) small, nevertheless it occurs. The 'small probability' in this definition corresponds with a sense of surprise, which we usually associate with the notion of chance, but on second thought it turns out that we are inclined to call thus every event whose occurrence is not certain in advance, that is to say, an event to which we attribute a less-than-one probability. The moment we adopt these intuitions as the initial working definition of chance, at once there arises the question what is meant by a (very) small probability or a probability of less than one. It appears that the only thing we can do under the circumstances is to directly invoke mathematical probability theory.

As we already know, (cf. Chapter 6), the contemporary standard form of probability theory developed by Andrey Kolmogorov constitutes a special case of the mathematical measure theory. As we remember (cf. Chapter 6.9), measure space is the triple (X, S, m) where X is a certain set, S – a family of its subsets, and m – the function that ascribes to subsets belonging to S non-negative real numbers. At the same time, certain appropriate axioms must be fulfilled, which we will refrain from repeating. A simple example of such a structure is the Euclidean plane X if we attribute to its subsets (e.g. to different geometrical figures) a number that constitutes the measure of their area. Additionally, if we assume that the measure of the entire set X equals one, i.e. $m(X) = 1$, we call function m a probabilistic measure, and the triple (X, S, m) – a probability space. On this assumption, the probability measure – probability, for short – of every subset X must fall between zero and one, i.e. be contained in the interval $[0, 1]$. Thus, a small probability of a certain subset means that its corresponding fraction is suitably small.

The mathematical probability theory is a purely formal theory that does not contain anything that would correspond to the sense of uncertainty or expectation usually associated with probability, or the sense of surprise or coincidence which we often associate with the notion of chance. These intuitions should be sought in the interpretations of probability calculus.

There are two distinct groups of interpretations of probability calculus, let us call them subjective (or epistemic) and objective (or ontological, cf. Chapter 6.7–6.8) interpretations, respectively. According to the first, probability constitutes a measure of our ignorance about the actual state of affairs, whereas according to the other, probability is a measure of indeterminacy of a certain event or a sequence of events that actually occur in nature. These two groups of interpretations suggest themselves so readily that Ian Hacking writes outright about a duality inherent in probability calculus, reminiscent of two interpretative polar opposites – epistemic and ontological ones.[106] Below, we shall deal with them in brief.

3. Subjective sources of probabilities

We are waiting at a stop wondering which tram will come first: number four or number eight? What is the probability of number eight coming first? Since both trams are already on their way, it is objectively known which one will come first, but we do not know this and therefore it may be legitimate to invoke probability calculus in order to settle the matter. In order to help us in the estimation of probability, we can draw on previous experience which suggests that we usually encounter twice as many number fours as number eights. This is a sample situation in which the application of the subjective interpretation comes naturally.

The subjective interpretation was developed by Bruno de Finetti who drew on the conception formulated by Thomas Bayes, the 18th-century clergyman and mathematician (cf. Chapter 6.8). This interpretation is based on the distinction between propositions and probabilities. A proposition can be true or false (subject to zero-one logic), however, probability is a measure (in Kolmogorov's theoretical sense) of the degree of somebody's belief in the truth of a certain proposition. The assignment of a measure representing the degree of certainty to a proposition is neither true nor false. The same proposition can be assigned different measures by different agents.[107]

---

[106] I. Hacking, *The Emergence... op. cit.*, pp. 11–17.
[107] The word 'agent' is understood here in its technical sense. It should not be construed to represent the man in the street but the man who constitutes part of the formal interpretation of the theory.

This assignment of the measure of the degree of conviction concerning a certain proposition is called its prior probability. The degree of conviction may change as new arguments appear. This measure is calculated by means of a formula proposed by Bayes. Prior probability is subjective, but all the rest, i.e. operations on this probability in compliance with the rules of probability calculus, is objective in nature.

The subjective interpretation is successfully applied to the discussion of observational errors. In measurements, we often face statistical or systematic errors. Statistical errors decrease as the number of measurements increases. These are successfully managed by the frequency (objective) interpretation of probability (Chapter 6.7). In this interpretation, probability is understood as the ratio of the number of 'favourable events' to the number of all possible events (on the assumption that all events are equally probable). There are several ways of coping with systematic errors that is those that arise as, for example, a consequence of effects outside the experimenter's control. Practice indicates that the subjective interpretation of probability is especially effective in this context. We start by defining the prior probability that draws on the information that we have on the present state of the measured system. As new information emerges, probability is modified in compliance with the principles of subjective interpretation. This strategy turns out to be effective in numerous measurements in the area of particle physics.[108]

Chance in the subjective sense occurs when an agent assigns a small degree of conviction (or anything less than certainty) to a certain proposition but, nonetheless, the event postulated by the proposition does actually occur. In this sense, what one agent may perceive as chance, it may not be so for another.[109] Chance thus understood

---

[108] Cf. e.g. G. D'Agostini, *Probability and Measurement Uncertainty in Physics – A Bayesian Primer*, arXiv:hep-ph/9512295.

[109] Subjective interpretations of probability are elaborated in more detail e.g. in the introductory parts of the article: C. M. Caves, C.A. Fuchs, R. Schack, *Subjective Probability and Quantum Mechanics*, "Studies in History and Philosophy of Modern Physics" 2007, 38, pp. 255–274.

constitutes an interesting subject of research on rationality of convictions, epistemology etc., but it is also an irremovable element of science, which, in effect, represents the incessant struggle of our ignorance against the Universe's resistance to reveal the secrets of its structure. However, is probability entirely reducible to a game between our current ignorance and the opportunities that open up ahead of us?

4. Objective sources of probabilities

Thus can probability be reduced exclusively to our ignorance? Physics prompts us to consider us yet another option when the application of probability calculus is imposed by the structure of the world without a significant share of our ignorance.

Physics investigates systems evolving over time, called dynamical systems. Such systems are called deterministic if their states at a certain point in time unequivocally determine their states at subsequent points in time. Studies of such systems do not require probability calculus (unless we need to take into account an inevitable element of our ignorance, e.g. when discussing measurement errors). A system that does not possess such a property is an indeterministic one. We cannot predict its future behaviour without referring to probability understood one way or another. Such systems are also called stochastic ones. Various possible evolutions of these systems are assigned different probability distributions, yet this does not mean that in stochastic systems anything can happen without any governing principles or constraints (as if we intuitively imagined 'the reign of chance'). The very possibility of application of probability calculus presupposes that the 'distribution of possibilities' must conform to its rules.

An interesting example comes from quantum mechanics. The Schrödinger equation describes the evolution of a quantum system (e.g. a subatomic particle). The equation is deterministic in the sense that the present state of the wave function unequivocally determines its future evolution. However, given the fact that the square of the module of the wave function is interpreted as the probability distribution of the fact that while measuring a certain quantum parameter (e.g. the location of a particle), we will find it in a certain value interval, we can say that the probability distribution at a given point in time unequivocally determines the probability distributions of various other states in the future, but the specific results of measurements are not determined. Consequently, we can say that the result of the measurement is random, but this randomness has its own well defined (although not in a deterministic way) place in the structure governed by the laws of quantum mechanics.

The theoretical situation described above translates into experimental results. A deterministic evolution of probabilities unequivocally predicts that one-half of the tritium atoms in a given sample will disintegrate within 4,499 days (this expectation has been experimentally confirmed to an accuracy of one to ten thousand), but exactly which atoms of the said sample will disintegrate within the nearest hour remains completely indeterminate.[110]

This situation (both theoretical and empirical) suggests that quantum processes are based on certain irreducible stochastic (probabilistic) phenomena that do not arise from our ignorance, but constitute a consequence of the structure of our world as it is. Naturally, there are interpretations of quantum mechanics that, notwithstanding, shift all the responsibility for the application of probabilistics at the fundamental quantum level to our ignorance (the so-called hidden parameter interpretations), but we know the interpretative price that must be paid anyway in the form of various quantum nonlocalities.

---

[110] Cf. T. Maudin, *What Could Be Objective about Probabilities*, "Studies in History and Philosophy of Modern Physics" 2007, 38, pp. 275–291 for an analysis of the objective interpretation of probability.

5. Instability versus ignorance

Let us take look at a situation that everyone would unquestionably qualify as an accident: a slightly misplaced railway switch due to a mistake made by a train signalman. As a result, two trains crash, causing numerous casualties. Putting aside the human dimension of this disaster, from the vantage point of physics, the situation can be described in terms of a relatively minor 'variation' that causes a substantial alteration in the trajectory of the system. This example was given by Maxwell.[111] Today, we treat it as an illustration referring to the so-called dynamical chaos. The theory of this phenomenon is at present a vigorously developing domain of physics and mathematics (theory of dynamical systems). As we have seen, Maxwell was aware of the existence of chaotic phenomena, their mathematical theory was developed by Poincaré (without calling it 'chaos'), and the issue of chance in their light was meticulously analysed by Smoluchowski. Let us briefly outline his conception.

Smoluchowski analysed the notion of chance in two of his works: *Uwagi o pojęciu przypadku w zjawiskach fizycznych*[112] and *Über den Begriff des Zuffals und der Ursprung der Warscheinlichkeitgesetzen in der Physik.*[113] A comparison of both these works proves that Smoluchowski struggled with the problem of chance and his views in this respect underwent a certain evolution.[114]

---

[111] J.C. Maxwell, *Matter and Motion*, Society for Promoting Christian Knowledge, New York, The Macmillan Co. 1920, pp. 13-14.

[112] *Remarks on the Concept of Chance in Physical Phenomena* in *Księga Pamiątkowa ku czci Bolesława Orzechowicza*, Towarzystwo dla Popierania Nauki Polskiej, Lwów 1916, pp. 445-458.

[113] "Die Naturwissenchaften" 1918, 17, pp. 253–263. The article was translated into Polish and published in "Wiadomości Matematyczne" 1923, 27, pp. 27–52.

[114] More information on the subject can be found in P. Polak, *Koncepcja przypadku w pismach Smoluchowskiego (The Concept of chance in Smoluchowski's writings)*, in *Krakowska filozofia przyrody w okresie międzywojennym (The Krakow Philosophy of Nature Between the World Wars)*, vol. II, eds. M. Heller, J. Mączka, P. Polak, M. Szczerbińska-Polak, OBI – Biblos, Kraków – Tarnów, pp. 443–460.

Smoluchowski treated probability calculus as a method for in-vestigating "events whose occurrence depends on chance."[115] The is-sue of chance in physics appears in the theory of errors that accom-pany each and every instance of measurement. Yet since the source of errors in measurement is ultimately to be found in our ignorance, Smoluchowski left out this problem. He was interested only in an objective event – such that is inseparably tied with the very 'core of physics.' Here is where Smoluchowski faced a problem: how could chance appear in an otherwise deterministic physics? (Smoluchows-ki died in 1916, when quantum mechanics was yet to become a rec-ognised paradigm). The final solution proposed by Smoluchowski is close to Poincaré's analyses. Smoluchowski considers chance to be a particular case of a causal relationship. It occurs when 'a small cause' brings about 'a large effect' or, using more contemporary language, when a small perturbation of the initial conditions causes substantial differences in future system trajectories. Smoluchowski's original observation consisted in that there must occur a dense compaction of initial states leading to extremely diverse results. Or to express it in more technical terms, the space of initial conditions must be charac-terised by an appropriate topology so that nearby states (in this topol-ogy) lead to distant states in the future. Smoluchowski thought that this was the only understanding of chance in the area of physics that was independent of the state of our knowledge since it was a feature of physical systems themselves.

However, the matter turns out to be a little more complicated, since what is meant by 'a small perturbation in the initial conditions?' In numerous local processes we can actually introduce perturbations in their initial conditions. We can, for example, throw a stone with different initial velocities and from different places. But how to cause perturbations in the initial conditions of the Universe's evolution? Our ignorance proves to be as important as a physical perturbation and this is related to every choice of initial conditions. Wishing to

---

[115] "Wiadomości Matematyczne" 1923, 27, p. 29.

preserve realism, we must take into account the inaccuracy that corresponds to inevitable measurement errors. And if even the smallest 'box of errors' (within which we choose the initial conditions) contains initial conditions that lead to 'divergent' trajectories, then the process is chaotic and if we wish to predict its states in the future, we must bring probability calculus into play.

Thus, chance as understood by Smoluchowski hinges on two factors: (1) on a certain physical property of a given system called its instability, which is characterised by the fact that a small change in the initial conditions (chosen on the initial space with an appropriate topology) leads to a great change in the future behaviour of the system); (2) on the fact that our awareness of initial conditions is always burdened with a certain error. The former has an objective significance, independent of our knowledge. The latter is subjective, although inherently related to the methodology of physics (measurement errors are unavoidable). In this understanding of chance, the objective and subjective factors, respectively, are not independent, but constitute elements of the same game of chance and determination.

6. The strategy of nature

Thus, there are two ways of understanding chance, or better still, two classes of understanding chance (since within every class it is possible to recognise more 'sub-understandings'): (1) chance in a subjective sense: when an event occurs whose probability is less than one in the subjective sense, and (2) chance in an objective sense: when an event occurs whose probability is less than one in the objective sense. Both kinds of understanding of chance often appear in our contacts with the world. Yet this should not be interpreted to mean that random events in a way destroy 'the order of nature.' On the contrary, nature quite often applies a strategy in which laws and chance cooperate. If we consider the operation of the laws of nature as a necessity, then we can say that necessity and chance cooperate in nature.

Let us consider another straightforward example. Let us try to put a sharpened pencil upright on the smooth surface of a table. The laws of physics tells us that we are attempting to achieve is an unstable state. If we cease to hold the pencil, it will doubtless fall. But which way? It depends on a number of 'random events': the fine twitches of my finger muscles, motions of the surrounding air, vibrations caused by a tram passing in a nearby street etc. Many of those random factors are random events only 'from the point of view' of the law of static of interest to us, but may equally well result from a non-accidental operation of other laws of physics. What is important, however, is that the law of nature in question leaves 'free room' for those 'random events' to act and there is exactly as much of this 'free room' left as necessary. Moreover, without their operation, the pencil in question 'would not know' which way to fall and the law could not operate.

And now, a more accurate formulation: the laws of nature are usually expressed via differential equations (or systems of such equations). Their solutions determine the possible behaviours of a given physical system, but a specific behaviour (a concrete phase-space trajectory) is dictated by the initial conditions or boundary conditions that must 'be imposed,' and thus are 'accidental' from the vantage point of the law of physics under consideration, though they do not have to be accidental from the vantage point of other laws of physics.

Consequently, we have arrived at a third kind of understanding of chance: chance with respect to a given law of physics, which may be the result of the quite non-accidental operation of other laws of physics.

It is an obvious thing (although we do not think about it frequently) that the laws of physics do not operate independently of one another, but combine to form a comprehensive (dynamic) structure in which the necessary (nomological) elements interact with the random ones. What we call 'a given law of physics' is only a certain aspect of such a structure. This aspect has been isolated by us from the rest not necessarily because it constitutes its naturally prominent part, but mainly by virtue of certain 'incidental' historical reasons.

7. Chance and sufficient reasons

The human mind has a certain very interesting property: we are inclined to consider a high probability of occurrence of a certain event to be a sufficient reason (in a sense close to Leibniz's understanding of the term) for its occurrence, whereas a low probability of occurrence of an event requires – in our opinion – a justification from another source. If such a justification cannot be found, we consider such an event to be random and are often surprised that it has actually 'happened.' This property of the human mind found its expression in the scholastic methodology in which the justification *ex communiter contingentibus* (by what commonly occurs) was considered valid in principle, though its only served to increase the validity of reasoning.

It may be legitimately supposed that this property of the human mind has evolved by way of accumulation of selection effects in the course of our biological history, that is – in other words – in the process of interaction with the surrounding world. The feature of the world that imposes these and not other behaviours, is its frequency stability, that is, a property of the world thanks to which the longer the sequence of random events, the closer the frequency of occurrence of a certain event in this sequence approaches a certain number. Let us emphasise, this is a property of the world, not a theorem of the mathematical probability theory. If the world did not have this property, the strategy of adaptation to events that occur frequently would bring about no evolutionary benefit. The occurrence of an event characterised by a low probability, that is, a random event, surprises us and demands justification, because it has no evolutionary 'justification.'

Probability calculus has yet another surprising feature, namely it applies to numerous classes of events which are strictly determined and with respect to which 'genuine randomness' is out of the question. In his pioneering work, Kolmogorov noticed that probabilistic methods could apply to the study of the distribution of occurrence of different digits (or sets of digits) in decimal developments of different numbers. This conjecture was fully corroborated. One may e.g. investigate the probability of occurrence of the sequence of digits 0123456789 in the decimal development of the number 'Pi.' Probabilistic methods are also used to investigate the distribution of prime numbers in the set of natural numbers (or zeros of the Riemann zeta function).[116] After all, the distribution of numbers on a real line or in decimal developments of numbers is determined with 'mathematical precision.' Is this, therefore, a typical example of probability in the subjective sense, whose source is to be found only in our ignorance?

A seemingly comparable situation affects the phenomenon of dynamical chaos. The classical process in which this phenomenon appears is a strictly deterministic process. Only our imprecise awareness of the initial conditions (due to the unavoidable measurement errors) causes that if we wish to find out about the future behaviour of a given system, we must apply probabilistic methods. However, the resemblance to the previously considered cases of distribution of numbers is not complete. In the case of dynamic chaos, we can argue that success in the application of probabilistic methods is guaranteed by certain structural properties of dynamic chaos, namely an appropriate topology of the space of initial conditions and the propagation properties of fluctuations of solutions. As far as we know, in the case of the distribution of prime numbers or decimal expansion, no analogy exists, unless our knowledge is incomplete in this respect as well. Yet one thing is certain, even the simplest mathematical theory, num-

---

[116] Cf. K. Maślanka, *Liczba i kwant (Number and quantum, in Polish)*, OBI, Kraków 2004.

ber theory, appears to conceal numerous profound mysteries, accordingly, what can we say about such an advanced mathematical concept as probability theory?

8. Concluding remarks

An advantage of defining chance as an event that occurs even though its probability is less than one is that the notion of chance can be entirely eliminated from physics and replaced simply with a probabilistic measure corresponding to the event under consideration. This cannot be done in philosophical considerations, where fuzzy qualitative concepts are unavoidable.

Yet despite the elimination of the fuzzy concept of chance from physics, further problems arise. First of all, today we know that there is more than one single probability calculus. Which one is used by nature for its own ends? Theoretical physics teaches us that in different physical theories different probability calculi should be applied and adapted to the mathematical structure of a given theory. In classical physics it is the classical probability calculus, but quantum mechanics requires a different understanding of probability.

The problem consists in that probability calculus, just as every good mathematical theory does, admits different generalisations. In the next chapter, we will review several such generalisations and it is not at all *a priori* obvious which plays a fundamental role in physics. At any rate, when we think about philosophical aspects of probabilistics, we cannot limit ourselves only to the classical probability calculus. Consequently, we must cope with a host of new questions. Some of them are listed below:

- Is the basic level of physics inherently probabilistic, and if so, which probability calculus governs it?

- Expressing the previous question in a more philosophical language, is 'the basic ontology' of the world probabilistic? In what sense?

- What is the role of probability in the macroscopic world? Is it only a tool that serves to fill the unavoidable gaps in our knowledge or – at least in some contexts – a trace of probabilistic properties of deeper levels of the physical world?

- And finally, what is the role of random events in our understanding of the world?

Chapter 8
Generalised probability calculi

1. Introduction

When contemplating the notion of probability, it is next to impossible not to devote more attention to quantum mechanics. Not only because it is *par excellence* a probabilistic theory, but also – or even first of all – because it has revolutionised our understanding of probability, although one must admit that the latter fact did not penetrate sufficiently deeply the awareness of numerous thinkers. Naturally, all those who have came across quantum mechanics know that it permits us to anticipate the results of experiments only with a certain degree of probability, but only those who have probed the mathematical apparatus of this beautiful physical theory in sufficient depth may realise the extent to which the classical notion of probability had to adapt itself to the requirements of quantum physics; and even they are not always conscious of the profound philosophical import of this fact. In this chapter I would like to focus on this particular issue. Yet not exclusively, since quantum mechanics not only revealed a new face of probability, but also initiated the process of its generalisation. Such generalisations, in turn, entail subsequent conceptual transformations. However, if what is subject to transformation is the very notion of probability, such transformations may force us to reconfigure a number of important philosophical concepts.

2. A bit of history

Let us begin with a historical introduction. As we remember (Chapter 6.3), Hilbert's address at the International Congress of Mathematicians in Paris in 1900 had a huge impact on the ensuing development of probability theory. Item six on Hilbert's list of unsolved mathematical problems pointed to the need to axiomatise those branches of physics "in which already today mathematics plays an important part." Such branches, in his opinion, were probability calculus[117] and statistical mechanics. Various mathematicians took up his challenge. As we know, in this line of research, Kolmogorov was the one to have finally axiomatised probability calculus.

In the autumn term of 1926–1927 in Göttingen, Hilbert delivered a series of lectures devoted to quantum mechanics, which was then still shrouded in a veil of thrilling novelty. He prepared these lectures with the help of his assistants Lothar Nordheim and Johann (John) von Neumann. Their collective monograph devoted to Hilbert's lectures[118] appeared in print soon afterwards. Von Neumann took a deeper interest in this problem area and before long he published three papers which turned out to be fundamental in this domain.[119] They gave rise to the crucial monograph devoted to the mathematical foundations of quantum mechanics.[120]

---

[117] At that time, probability calculus was virtually identified with its applications in physics.

[118] *Über die Grundlagen der Quantenmechanik*, "Mathematische Annalen" 1927, 98, pp. 1–30.

[119] *Mathematische Begründung der Quantenmechanik*, "Göttinegen Nachrichten" 1927, pp. 1–5; *Warscheinlichkeitstheoretischer Aufbau der Quantenmechanik*, "Göttinegen Nachrichten" 1927, pp. 245–272; *Thermodynamik quantenmechanischer Gesamtheiten*, "Göttinegen Nachrichten" 1927, pp. 273–291. These works can also be found in J. von Neumann, Collected Works, vols. I–III, A.H. Taub, ed., Pergamon Press, New York – Oxford 1961.

[120] J. von Neumann, *Mathematische Grundlagen der Quantenmechanik*, Springer, Berlin 1932.

In von Neumann's approach, the basic role was played by operator algebras on Hilbert spaces. The well-known work by Murray and von Neumann[121] lay the groundwork for the now elaborate theory of these operators. A certain class of those, which attracted the attention of Murray and von Neumann, over time became known as von Neumann algebras. It is this very class of operators that plays a fundamental role in the theoretical layer of quantum mechanics. A definition of von Neumann algebras would require much more advanced technical resources, but let us review at least some of their (amazing!) features.

3. A bit of mathematics: some properties of von Neumann algebras

In the mathematical formalisation of quantum mechanics, the states of a quantum system (e.g. an electron) are represented by elements (also called vectors) of a Hilbert space. The basic structure of a Hilbert space is the same as that of ordinary vector spaces, but this structure has a superadded rich layer of properties thanks to which Hilbert spaces perfectly model the states of quantum systems.[122]

If we measure a certain property of a quantum system, e.g. the electron spin, the very act of measurement perturbs the system and modifies its state. This is how quantum measurements differ from macroscopic measurements. By measuring the length of a table with a rod, we do not perturb the measured quantity in any perceptible way. Conversely, measurement 'on a quantum system' (as physicists like to say) constitutes a veritable invasion, which

---

[121] F.J. Murray, J. von Neumann, *On Rings of Operators*, "Annals of Mathematics" 1936, 37, pp. 116–229.

[122] A fairly accessible presentation of the mathematical structure of quantum mechanics is contained in part two of my book *Some Mathematical Physics for Philosophers*, Pontifical Council for Culture, Pontifical Gregorian University, Rome 2005.

results in the system passing from its 'initial' to 'final state.' This process is accompanied by obtaining a certain number or numbers (readouts) on the dial of the measuring instrument (or in appropriate computer inputs). Exactly the same occurs if we use an appropriate operator on a vector in a Hilbert space that represents the state of a certain quantum system. This operation causes the 'initial vector' to change into the 'final vector' of the Hilbert space, with the operator producing numbers as possible measurement results.

Is it not amazing that actual quantum processes obey some abstract mathematical operations?!

We all know the operation of projecting. The sun projects my shadow onto the flat pavement. Any secondary school student knows (or should know) how to project a vector onto an axis of the reference frame. Vectors in a Hilbert space can also be projected onto different directions or planes (more generally, onto different subspaces). Thus, we have a vector before projecting onto a certain subspace of a Hilbert space, a vector after projection onto this subspace and the projection operations. Let W be a (linear) subspace in Hilbert space H, and ψ a vector in Hilbert space H. This vector can be projected onto subspace W. Let us express this operation as follows:

$$P_W : \psi \to \psi_W ,$$

where ψ_W is a vector after its projection onto subspace W. Operation P_W is called a projection operator (onto subspace W). Let us now consider subspace W_{ort} that is orthogonal ('perpendicular') to subspace W.[123] Let us assume that vector ψ that we want to project onto subspace W already belongs to subspace W, so $\psi = \psi_W$. As a result of projecting (using operator P_W on ψ_W) we will naturally obtain the same vector ψ_W, so

---

[123] The notion of orthogonality is well defined for every Hilbert space.

$$P_W : \psi_W \rightarrow \psi_W .$$

However, if vector ψ belongs to subspace W_{ort} that is orthogonal to W, which means that $\psi = \psi_{ort}$, as a result of acting with operator P_W on this vector, we obtain a null vector, or

$$P_W : \psi_{ort} \rightarrow \mathbf{0} .$$

Putting this property into the language of mathematics, we say that projection operators have only two eigenvalues: one (1) if the projection reproduces the projected vector, or zero (0) if the projection results in a null vector.[124]

Let us now consider a certain operator algebra on a Hilbert space. If it is a von Neumann algebra, then it can be reconstructed entirely from projection operators (in mathematicians' language, such an algebra is called 'generated by projection operators'). One may also view it from a somewhat different angle: instead of talking about projection operators, we can talk about the subspaces on which these operators project. From this perspective, von Neumann algebra can be completely recovered by examining the structure constituted by all the subspaces of a Hilbert space associated with a given operator algebra under consideration (i.e. Hilbert space on which these operators are acting).[125] This very point of view was liberally applied by Murray and von Neumann in their original work, with other mathematicians following in their footsteps.

---

[124] Please note that it is an example, not a definition, which can be found in any textbook on quantum mechanics.

[125] These 'projectional properties' of von Neumann algebra can be used to define this algebra. There is another, purely algebraic way of defining von Neumann algebra (without a reference to a Hilbert space). In this approach, von Neumann algebra is identified with a dual space with respect to a certain Banach space. The latter is unique up to isomorphism.

4. Von Neumann's agenda

Another International Congress of Mathematicians was scheduled to take place in Amsterdam in 1954. The organisers asked von Neumann to formulate – much like Hilbert had done half a century before – a list of the most important unsolved problems in mathematics. Von Neumann not only accepted the challenge, but also to a considerable extent pursued the path charted out by his predecessor. In his Sixth Problem, Hilbert postulated the axiomatisation of those branches of physics "in which mathematics plays a dominant role." At the time, quantum mechanics had become such a branch of physics, so von Neumann suggested that it should be axiomatised as well. Besides, his work to date clearly followed this direction. His agenda was ambitious. He wanted to create an axiomatic system in which logic, probabilistics and quantum mechanics would acquire their geometric interpretation. As result, quantum mechanics together with its deeply probabilistic properties would gain a logical transparency and geometric precision.

Von Neumann's agenda did not gain such renown as Hilbert's famous "Unsolved Problems in Mathematics." It may have been premature. Today we know that before attempting to unify logic, geometry and probability theory, all these disciplines need to be radically generalised. As we will see below, this process is already underway. However, in order to grasp von Neumann's intuitions, let us take yet another look at the operator algebra on a Hilbert space.[126]

As we know, projection operators on vectors in a Hilbert space produce only two eigenvalues: either one or zero. Here emerges the association with logic. As we know, in propositional calculus (propositional logic), true propositions are assigned the logical value of one (1), whereas false propositions are assigned zero value (0). It turns out that by means of projection operators we can reproduce the en-

---

[126] A more exhaustive treatment of the subject can be found in G. Valente, *John von Neumann's Mathematical 'Utopia' in Quantum Theory*, "Studies in History and Philosophy of Modern Physics" 2008, 39, pp. 860–871.

tire classical propositional calculus, but with an important exception: the 'operator logic' does not meet the classical law of distributivity of conjunction with respect to alternative.[127] This reflects the specificity of quantum mechanics where – as a result of its probabilistic character – the difference between 'and' (conjunction) and 'or' (alternative) becomes fuzzy.

Since the structure of projection operators is impressed on the structure of a Hilbert space, it is fair to say that the structure of a Hilbert space (due to the geometry of its subspaces) encodes the logical propositional calculus, although modified in comparison with the classical propositional calculus, moreover, it is modified to an extent required by quantum mechanics.

The relationships between logic and further development of probability calculus reach considerably deeper than we have the opportunity to present here.

5. A bit of probability calculus

Once again, let us bear in mind the definition of a probability space (Chapter 6.9). It is a triple (X, S, p) where X is a certain set (we will also call it a space), S is a set of its subsets,[128] and p is a function that to subsets belonging to S attributes numbers from the interval $[0, 1]$, which we note as $p: S \rightarrow [0, 1]$. Subsets that belong to S are called events, whereas the numbers $p(s)$, $s \in S$, are called the probability that event s may occur, and the fairly obvious equality holds:

$$p(s_1 + s_2 + s_3 + \cdots) = p(s_1) + p(s_2) + p(s_3) + \cdots,$$

which says that the probability of the sum of events equals the sum of probabilities of those events (for any countable set of disjoint events).

---

[127] In a standard logical notation it is expressed thus: [p ∧ (q ∨ r)] <=> [(p ∧ q) ∨ (p ∧ r)].
[128] That fulfils the axioms of a σ-algebra.

An important notion in probability calculus is that of the random variable. This is a function that assigns numbers to events. In daily life, we also sometimes assign numbers to certain objects or to events, e.g. we may assign the measure of length to a table, or to the sunset, the time at which it occurs (for the sake of simplification, we will also use the term 'events' when referring to objects). Random events cannot be unequivocally assigned such 'measures,' one can however assign to them numbers in accordance with a certain established principle so that they can somehow be compared with one another, e.g. the results of coin tosses can be assigned 1/2 each, and each throw of a die can be assigned 1/6. This procedure enables us to transfer certain regularities that apply to real numbers onto the occurrence of random events. Formalising these intuitions, we can say that the random variable is function f defined on a set X that assumes its values in the set of real numbers $\mathbf{R}$, thus $f: X \to \mathbf{R}$, with an appropriate axiom ensuring that the function transfers certain 'good properties' of the space of real numbers onto space X.[129]

Let us consider any subset A of space X that belongs to the family of subsets S and let us define function χ_A in the following way: $\chi_A(x) = 1$ if x belongs to set A, and $\chi_A(x) = 0$, if x does not belong to set A. It is obvious that function χ_A defines set A in an unequivocal manner – we call it the characteristic function of set A. Let us denote the set of all functions of this kind (for all subsets that belong to family S) as $\mathcal{P}(S)$. We thus arrive at an interesting conclusion. It turns out that these functions generate a certain von Neumann algebra, in which they play the role of projection operators. As befits projection operators, they produce numbers: one and zero. The von Neumann algebra under discussion consists of functions defined on probability space (X, S, p). It is denoted by the symbol $L^\infty(X, S, p)$.[130] Since this particular von Neumann algebra consists of functions, and the product of

---

[129] The axiom that guarantees this property is as follows: If B is an element of Borel σ-algebra on $\mathbf{R}$, then $f^{-1}(B)$ is an element of S, thus S is a Borel σ-algebra on X.

[130] Spaces of this kind are well known in mathematics. They consist of measurable bounded functions.

a multiplication of functions does not depend on their order, we are dealing with a commutative von Neumann algebra. But such an algebra should act on a Hilbert space. This is the case this time as well. The Hilbert space on which the von Neumann algebra $L^\infty(X, S, p)$ acts is denoted by $L^2(X, S, p)$. This is also a functional space.[131] Functions that belong to von Neumann algebra of the type $L^\infty(X, S, p)$ act on the functions that belong to the Hilbert space $L^2(X, S, p)$ by ordinary multiplication.

In this way, we have arrived at an interesting conclusion: the entire classical probability theory comes down to the existence of a commutative von Neumann algebra $L^\infty(X, S, p)$ generated by characteristic functions (projections) that belong to $\mathcal{P}(S)$ that acts on the Hilbert space $L^2(X, S, p)$. We must now answer yet another question, namely what corresponds with the probabilistic measure p in this construction. This requires a few words of introduction.

Let us consider an algebra, e.g. von Neumann algebra $M = L^\infty(X, S, p)$. The (linear) mapping φ that assigns real or complex numbers to elements of this algebra, i.e. $\varphi : M \to \mathbf{K}$ (where $\mathbf{K} = \mathbf{R}$ or $\mathbf{K} = \mathbf{C}$) is called a functional on this algebra.[132] If such a functional φ has two additional properties: (1) it is positive[133] and (2) 'normed to unity,'[134], we call it a state on algebra M.

It turns out that probabilistic measure unequivocally determines a certain state on von Neumann algebra M.[135] Now we can say in brief that the classical probability calculus is a pair (M, φ) where M is a commutative von Neumann algebra $L^\infty(X, S, p)$, and φ is a certain state on this algebra.[136]

---

[131] This functional space is also well known in mathematics. It is called the space of square integrable functions.

[132] In the case under consideration, it can be said that the functional is a 'function on functions.'

[133] I.e. $\varphi(f^*f) \geq 0$ for each f that belongs to M.

[134] Which means in the language of mathematics that the norm of the functional φ equals unity, $\|\varphi\| = 1$.

[135] This state is expressed by the following formula $\varphi(f) = \int_x f(x)dp(x)$.

[136] In this definition we need not mention the Hilbert space $L^2(X, S, p)$, because its existence is entailed by the von Neumann algebra $L^\infty(X, S, p)$.

Is this entire complex construction only a matter of mathematical elegance? Not only. The initial definition of probability space (X, S, p) is without a doubt more straightforward and easier to handle, but if we were to confine ourselves to it, there would be no prospect for further generalisations, while expressing probability theory in the language of commutative von Neumann algebras immediately suggests the way in which probability theory can be further generalised. Why do von Neumann algebras have to be commutative? What if we used noncommutative von Neumann algebras?

6. A noncommutative probability theory

The last question in the preceding subsection includes a suggestion: let us discard the assumption of the commutativeness of von Neumann algebra and assume that generalised probability space is represented by the pair (M, φ) where M is not necessarily a commutative von Neumann algebra and φ a state on M. This is a significant generalisation, because if we limit ourselves to a case where M is a commutative von Neumann algebra, we obtain the classical probability space.

In order to ensure that our generalisation is correct, we must impose a certain limitation on the state φ. The state φ, understood as a generalised probabilistic measure, must guarantee the fulfilment of the counterpart of the classical rule which states that the probability of the sum of events equals the sum of probabilities of these events. The state that meets this condition is called a normal state.[137]

In summary, we will say that the probability space is a pair (M, φ) such that M is a von Neumann algebra (not necessarily a commutative one), and φ a normal state on M. This is a strong gener-

---

[137] A state φ on von Neumann algebra M is a normal state if it fulfils the relationship: $\varphi(\sum_{n \in N} P_n) = \sum_{n \in N} \varphi(P_n)$ for each countable family $\{P_n\}_{n \in N}$ of mutually orthogonal projection operators in M.

alisation of the classical case, which is manifested, among other things, by the fact that in the classical case we have, in principle, only a single mathematically interesting probabilistic measure,[138] while in the noncommutative case there exists a wealth of different measures.

If the Reader has managed to get through this arduous chain of subsequent notions, he will surely notice a conclusion that immediately suggests itself: the notion of probability that operates in quantum mechanics is tied with the noncommutative probability theory. This is indeed the case. One may even say that all conceptual and interpretative difficulties in which quantum mechanics is involved result from the fact that the intuitively clear notion of classical probability was replaced in it by its noncommutative generalisation.

It does not mean, however, that the classical notion of probability has been completely eliminated from quantum mechanics. Not all operators that correspond to observable quantities multiply in a noncommutative manner – where noncommutativeness does not occur, all calculations are performed within the classical model. It does not destroy the consistency of the mathematical structure of quantum mechanics, after all, the classical probability theory is but a limiting case of the noncommutative theory. What we see here is a really beautiful mathematical structure.

To complete our reasoning, let us add that an important extension to the standard (non-relativistic) quantum mechanics is quantum statistics and the quantum field theory. They analyse systems with an infinite number of degrees of freedom. This requires a significant generalisation of the mathematical apparatus. Inasmuch as in the case of standard quantum mechanics one may still insist that the algebraic presentation is only a matter of greater elegance, in the case of quantum statistics and quantum field theories, the algebraic approach becomes an indispensable research tool. Moreover,

---

[138] Known as Lebesgue measure.

von Neumann algebras, necessary for this purpose, are characterised by a markedly greater degree of complexity and require a considerably greater mathematical artistry.[139]

7. Free probability calculus

An important notion in the classical probability theory is that of the independence of random events. We said, for example, that subsequent rolls of a die or tosses of a coin are not interdependent, i.e. the result of a given roll/toss does not depend on the results of previous rolls/tosses. This assumption plays a crucial role in numerous theoretical considerations and in various practical applications. This notion also has its own noncommutative generalisation.

But in generalisations of various mathematical structures and different mathematical theories, there sometimes appear notions that do not have their counterparts in structures or theories which have given rise to a given generalisation. For example, the spin of an elementary particle, a very important notion in quantum mechanics, has no counterpart in classical physics. It turns out that in the noncommutative probability theory there exists a concept similar to independence, however – strictly speaking – it has no its counterpart in classical probability. The notion originated in the mid-1980s with Don Voiculescu who worked on certain issues related to von Neumann algebras. The subalgebras of von Neumann algebras, 'independent' in this new sense, were called by Voiculescu 'free subalgebras.' Over time, this current of research gave rise to a new branch of probabilistics nowadays called the free probability theory.[140] More and more as-

---

[139] In principle, nonrelativistic quantum mechanics employs only von Neumann algebras of type I, while in quantum statistics and quantum field theories von Neumann algebras of all types are used.

[140] A good introductory article on the subject was written by Ph. Biane, *Free Probability for Probabilists*, arXiv: math.PR/98/9809193. Another useful reference is: D.V. Voiculescu, K.J. Dukema, A. Nica, *Free Random Variables*, American Mathematical Society, Providence 1992.

sociations of this theory with different branches of mathematics have been revealed; it is also more and more fruitfully applied in physics. In order to illustrate the surprising effectiveness of this new theory, let us consider a single but very distinctive example. The problem is purely mathematical, but it has important consequences for quantum physics.

In practical calculations, the elements of von Neumann algebras are represented by means of square matrices with N rows and N columns (the so-called $N \times N$ matrices). As we know, some matrices (operators)[141] represent measurable quantities. Such matrices produce numbers (the so-called eigenvalues) that represent the possible outcomes of measurement of a given observable quantity. A set of such numbers is called the spectrum of a given matrix (operator). Let there be given two such matrices, matrix A and matrix B, both of the $N \times N$ type. Let us further assume that we know their spectra and we wish to calculate the spectrum of the matrix $A + B$. For a large N, the task is very difficult to solve. This is precisely where the free probability theory comes to the rescue. It permits us to rigorously formulate and prove the following conclusion: if in the above-mentioned example the number of columns (and rows) in matrix A and B approaches infinity (i.e. $N \to \infty$), then we can calculate the spectrum of the matrix $A + B$ – even without knowing the exact structure of matrices A and B, respectively – by means of a specific formula[142] and the probability that the obtained result will be accurate is very high.

This conclusion is really surprising. 'Really large' values of N ($N \to \infty$) generate probabilistic properties that do not exist for small values of N. Nothing of the kind is present in the classical probability calculus.

---

[141] That is, Hermitian matrices (operators).

[142] If the probabilistic measures on the spectra of the matrices A and B are μA and μB, respectively, then the probabilistic measure on the spectrum of the matrix A + B is a free convolution of measures μA and μB.

8. Consequences

The existence of generalisations of the classical probability theory has
far-reaching consequences for mathematics, physics and for certain philosophical considerations.

Axiomatisation of the classical probability calculus accomplished
by Kolmogorov pushed probabilistic considerations into the mainstream
of the development of mathematics and, by the same token, triggered
its powerful interactions with other mathematical theories (cf. Chapter
6). The natural direction in the evolution of mathematics is to strive for
generalisations. The first sign of the need to produce a generalisation
of the classical notion of probability arrived from physics. As was often the case in the history of science, in the preliminary phase, quantum
physicists applied generalised probability in a spontaneous yet not quite
conscious manner, with mathematical specification arriving in its wake.
First, physicists followed the path of algebraisation of the classical probability theory (commutative von Neumann algebras) and then the obvious generalisation with the required applications to quantum physics
(noncommutative von Neumann algebras).

The process, once started, began to bear new fruit. The discovery of
free probability theory opened up completely new possibilities for probabilistics in that it drew attention to its close links with a new rapidly developing branch of mathematics – noncommutative geometry. This theory is currently revolutionising many traditional mathematical notions.
Many of its models are strongly non-local ones. Local notions – such as
the concept of a point or its neighbourhood – make no sense in them,
which entails further consequences. As Connes writes, "noncommutative sets are thus characterised by the effective indiscernibility of their
elements."[143]

---

[143] A. Connes, *Noncommutative Geometry*, Academic Press, New York 1994, p. 74.
What we understand by effective indiscernibility is the impossibility to distinguish
amongst the elements of a set by means of a countable family of their features. Such a
conclusion is obtained by Connes on the assumption that all mappings applied in the
reasoning are measurable. Rejection of this assumption would also constitute a serious
'pathology.'

Are such sets still sets in the traditional understanding of the term? Does this question not herald the arrival of a post-Cantorian mathematics? If such a revolution is indeed in the offing, noncommutative probability theory will be taking an active part in it.

All that occurs in mathematics will sooner or later resonate in physics. Applications of noncommutative geometry together with noncommutative probability theory are already taking place. A natural area of interest is the quest for a fundamental physical theory that would unify quantum physics and general relativity (combining them to form a quantum theory of gravitation) and a synthesis of all four basic physical forces: gravitation, electromagnetism, weak and strong nuclear forces. Most work in noncommutative mathematics carried out with physics in mind either directly refers to this programme or develops theoretical tools to cope with it. If it turns out that the fundamental level, the so-called Planck level, is indeed governed by noncommutative mathematical structures, the discovery of these structures will surprise physicists no less than the discovery of quantum mechanics surprised their friends from the first half of the 20th century. Today's disputes over whether the Planck level is fundamentally probabilistic, or else, governed by laws similar to classical physics, may prove to be just as naive as the 19th-century disputes concerning the existence or non-existence of atoms appear irrelevant when compared with today's advances of particle physics. If a certain version of noncommutative probability theory does apply at the fundamental level, we will have to alter considerable conceptual networks that have a crucial significance for our understanding of physics and the world in general. This conclusion will have to be additionally reinforced if noncommutative mathematical theories trigger yet another revolution in the foundations of mathematics. Changes in mathematical thinking will be reflected not only in the philosophy of physics, but will also have a general philosophical import.

However, in order to appreciate the philosophical significance of generalisations of the classical probability theory, it is not necessary to wait for a revolution in the foundations of mathematics and for physicists to develop their fundamental theory. Even now the noncommutative probability theory has something to offer to philosophers. Specifically, probability calculus should be treated just as all other mathematical theories. This is not a banal lesson, since all too often probability calculus (still in its classical version) is treated as a kind of 'superior ontology,' interpreting a high probability of occurrence of an event as its sufficient raison d'être. It is definitely justified in the area of proper application of the classical probability calculus, but the extension of this strategy onto the basic level of physics or onto certain metaphysical considerations may not be acceptable. For example, there once was a conception which stated that the very basic level was completely chaotic, that there were no regularities in it, with the laws of physics having emerged from this chaos as a result of averaging and probability games.[144] This conception assumed a general validity of the principles of the classical version of probability calculus which were to fulfil the role of 'superlaws' with respect to the ordinary laws of physics and thus account for their existence. Today, such absolutisation of the classical probability calculus is no longer allowed. Why should the classical probability calculus that works well on the macroscopic scale be better than its generalisations outside this field of application and be elevated to the proportions of 'an explanatory ontology'?

Another domain affected by the 'operation' of the principles of the classical probability calculus that endows it with the status of 'an ultimate explanation' is the 'set of all universes.' The very conception is highly controversial, and the manner in which probability calculus is applied within it makes it even more suspect. In speculations concerning 'the multiverse' we often come across such questions as: What is the probability of initial conditions such as those in our Universe becoming realised in one of the possible universes? What is the frequency of occurrence of carbon in the set of universes? What is the distribution of probability of consciousness occurring in the multiverse? What is the probability of occurrence of universes that contain a lot of black holes? Etc., etc..

---

[144] I write about this in more detail in *Filozofia i Wszechświat (Philosophy and the Universe)*, *op. cit.*, pp. 60–63.

Here again probability calculus plays a privileged part but which probability calculus? How is one to define the probabilistic measure on the set (and is this really a set?) of all possible universes?[145]

Probability theories, since they are many, should be treated as all other mathematical theories. All of them, together with other mathematical theories, combine to form an impressive superstructure which is amazing in both its harmony and architecture. I do not presume that any one of these theories has a distinct ontological significance, however, the fact that the structure of the Universe is characterised by such an effective correspondence with mathematical structures should constitute an object for deep philosophical reflection.

---

[145] I write about these issues more extensively in *Ultimate Explanations of the Unvierse*, Springer, Dordrecht 2009, pp. 110–112. An in-depth discussion on the possibility of introducing probabilistic measure on families of universes can be found in G.F.R. Ellis, *Multiverses: Description, Uniqueness and Testing*, in *Universe or Multiverse*, B. Carr, ed., Cambridge University Press, Cambridge 2007, pp. 387–409.

Chapter 9

Probability in the structure of the Universe

1. Probability – from applications to theory

It makes sense to begin the recapitulation of the second part of our reflections on probability and chance with a comparison between what happened in the area at the stage of 'beginnings and consolidations' (Part I) and what took place at the stage of 'maturity and further development' (Part II). In nature, the period of maturity is followed by a period of withering whereas in mathematics maturity always signifies the beginning of a new stage of dynamic development. One may go as far as to say that the early stages of probability calculus were more difficult than those of other branches of mathematics, such as arithmetic or geometry, since it was necessary not only to overcome the resistance of the mathematical matter characteristic of all kinds of early developments in this domain of science, but also to penetrate an additional barrier in the form of a persistent conviction that chance constitutes a breach in rationality and therefore it cannot be subjected to mathematicisation. We have seen how many centuries had to elapse in order to first soften, and afterwards completely neutralise this conviction (it still lingers on in popular views and in some philosophical doctrines).

However, in another respect, the first period of work on probability calculus bore all the hallmarks of the style that had characterised the other branches of mathematics. What I have in mind is that there was still no clear distinction between mathematics as a purely formal science and its application to research (modelling) of the world.

In principle, the difference between mathematics and its application was tacitly understood, but the demarcation line between them often meandered, and at times even became completely blurred. In probability calculus, this distinction was all the more difficult to apply. For example, in the context of games of chance, there was no apparent need for carefully designed experiments in order to graduate from dice rolling to mathematical analyses. Mathematics simply grew out of situations characterised by expectations and risks.

Mathematical self-awareness gradually became keener, but it was not until the second half of the 19th century that it resulted in the rapid growth of what was later called meta-mathematics and which contributed to the rapid progress of formal logic and formal aspects of the philosophy of science. But probability calculus was fairly slow to enter the formalisation stage. The mainstream of its development still occurred via its applications.

The fact that progress in the understanding of probability was related to the work of physicists such as Maxwell, Bolzmann, Gibbs or later Einstein and Smoluchowski, was quite understandable. Since its very beginnings, probabilistics had been linked with the investigation of random phenomena that occur in the world. What may be surprising is only the degree of sophistication of works the above-mentioned physicists – both in terms of mathematical subtleties and penetration into the physical mechanisms of the processes studied. The foundation of statistics as a branch of abstract mathematics and the institution of statistical physics went hand in hand – the two mutually conditioned one another.

However, the relationship between probability and deep arithmetic regularities appeared quite unexpectedly in astronomical applications. If the sequence of certain integers that appear in continued fractions used by astronomers to calculate the disturbances of planets' orbits is divergent, then, in general, the given orbit is unstable, with the probability of divergence of such a sequence is less than an arbitrarily selected number, in other words, it is 'very small.' But what is meant by a 'very small' probability? These at first somewhat vaguely formulated problems, turned out to be quite deep. Poincaré's insightful works concerning the three body problem (which afterwards led to the famous Poincaré recurrence theorem) and the innovative works by Borel in the area of mathematical measure theory were needed in order to rigorously formulate these issues and to achieve real progress.

2. Measure and probability

Measurement constitutes a basic physical operation. Physics is *par excellence* an experimental science; every experiment comes down to the measurement of a certain quantity. The performance of measurements constitutes a practical skill and quite often requires considerable expertise, however, mathematicians, as they do, with their propensity for abstraction, have extracted essential properties from measurement endeavours, converted them into axioms and created an entirely new branch of mathematics called measure theory. Moreover, soon it turned out that this is one of the basic mathematical theories in that a lot of its aspects are interwoven into the fundamental aspects of other branches of mathematics. Every measurement consists in the assignment to a certain 'object' of a number which constitutes its measure. After all, mathematics, 'gets by' assigning numbers to various objects. It is enough to realise how many mathematical operations end with obtaining a number. And if such treatment fulfils the axioms of the mathematical measure theory, it constitutes measurement in the mathematical sense.

Now, it was enough to notice (but noticing such 'straightforward' regularities often requires exceptional insight) that probability consists in the assignment of a number to a certain event (e.g. the number ½ to the event of obtaining tails). The fact that it is an event and not an object is irrelevant (for that very reason, the word 'object' in the preceding paragraph was put in inverted commas). Probability is thus a measure, and probability theory is a particular case of the mathematical measure theory. A particular case, because mathematical measure in order to be a probability, must meet one additional condition: it must be 'normalised to unity,' i.e. the probability that a certain event (from among all the possible ones) will occur equals one, for example, the probability of obtaining either a heads or a tails in a coin toss equals one.[146]

Although for some time the relationship between probability and measure had already been almost obvious for many mathematicians, a breakthrough in this area did not occur until the publication in 1933 of a seminal work by Andrey Kolmogorov. His case is very instructive. As was the case with many scientists before him, he was fascinated by the relationship between the mathematical notion of probability and the phenomena that occur in the physical world. What does it mean to say that a process is random? How do deterministic processes differ from stochastic processes? It is hard to imagine anyone occupied with probability theory who would not ask similar questions of himself. However, Kolmogorov was a proper mathematician and was able to clearly distinguish the physics of the real world from its mathematical models. A mathematical model is a purely logical construction and as such belongs to the domain of mathematics. This is the price that must be paid for accuracy but it is an accuracy which pays because it shows relationships amongst various concepts and, as a result, offers an insight into the architecture of mathematical structures. When afterwards these structures are applied in physics, the insight thus obtained transfers from mathematics to physics – and we understand better how the world works.

3. Interpretations

The whole of theoretical physics operates in this way. Mathematical structures, by means of which we model different aspects of the world, tend to have their own long-standing histories. They are born

---

[146] To put it somewhat more rigorously, an event is a subset of the space of outcomes. Each event S has its probability or measure, $p(S)$. The fact that this measure is normalised to unity means that the probability of occurrence of a certain event equals 1, whereas the probability of $p(S)$ of each and every event S is $0 \leq p(S) \leq 1$.

and change throughout arduous and sometimes extremely tangled processes of research into specific natural phenomena. We saw this when we discussed the gradual consolidation of probability calculus. However, as sometimes happens and especially so in recent times, a certain mathematical structure is created by a problem situation within mathematics itself. It happens when already well-known mathematical structures are subjected to evolution and impose the development of certain new constructions which, for one reason or another, turn out to be important for mathematics. Sometimes only later do they find a rewarding application to investigating the world.

In actual research, this process of interaction between mathematics and the investigated aspect of the world usually takes place in an instinctive manner – the investigator's creative intuition merely prompts him which mathematical techniques to apply to 'solve the problem at hand.' But at the 'mature stage' intuitions must be harnessed, everything must be organised, properly analysed and the borderlines of mathematics must be carefully established. It must be decided what the mathematical model of the world is and how it relates to the world itself. At this last stage, a crucial role is played by experimental procedures.

The fact that the mathematical structure created by Kolmogorov (measure theory normalised to unity) was precisely the structure that needed to be created, was promptly confirmed by an outright invasion of the said structure into other branches of mathematics. As I have already mentioned, today the structures of measure (apart from such structures as topological, algebraic and analytical ones) belong to the inventory of basic mathematical structures, with the most interesting things in contemporary mathematics taking place in areas where these structures interact with one another.

Let us look once again at probability theory as a measure theory normalised to unity. Let us note that neither mathematical measurement (i.e. the attribution of numbers to certain objects under certain conditions) nor the postulate that the measure must be normalised to unity, contains anything of the uncertainty or volatility of the expected result that we intuitively associate with probability. However, we would not wish to give up these intuitions all too easily since we believe that they capture a certain extremely important aspect of our experience, and that is how the problem of the interpretation of mathematical probability is born.

As we know, there are two groups of the interpretation of probability: the objective – those seeking the sources of random variation in the properties of nature independent of the cognitive subject, and the subjective – those holding that we are ignorant of the real reasons at work in the world which are responsible for random variation. Naturally, in certain contexts, both sources merge with each other. While mathematical probability theory (understood as a purely formal one) is, in principle, dependent on philosophy to the extent to which all mathematics sometimes constitutes the object of philosophical disputes, the problem of interpretation of probability is, from the very outset, deeply entangled in philosophical nuances.

Today, it would be hard to defend the view that our ignorance is the only source of randomness in the world. Even in classical mechanics there are numerous unstable processes that force us to use probabilistic methods to investigate them. Admittedly, at the basic level these processes are characterised by a rigorous determinism, i.e. our awareness of initial conditions with an infinite precision would permit us to unequivocally predict the behaviour of the system, but the problem is that infinite precision is unattainable to us. Still, is it not a matter of our ignorance? Surely, it makes no sense to quibble about words. At any rate, it is an objective feature of nature that, apart from unavoidable measurement errors, there exist various fluctuations that attack the system under consideration from outside, which – even in principle – make it impossible to know the initial conditions with any degree of accuracy.

The problem becomes more pronounced when we consider the microscopic world. It is not so much that quantum indeterminacies place tight constraints on our capacity to recognise the initial conditions. A considerably deeper source of inevitability of probabilistic methods is that important quantum equations (e.g. the Schrödinger equation) do not determine the evolution of certain properties of the system (e.g. positions and momentums of particles), but the evolution of their probability distributions.

As we know, there are interpretations of quantum mechanics with hidden variables which yield, in principle, the same empirical predictions as the standard version of quantum mechanics. Although these interpretations eliminate indeterminism, they still leave a strong non-locality of quantum systems and thus do not permit us to preserve all the 'intuitively obvious' properties of the macroscopic world.

What remains is the problem of the fundamental level. Is it essentially probabilistic, can it in some way 'vindicate' the more or less classical determinism? This question will be answered by the quantum theory of gravitation or some other theory of the fundamental level once we have discovered it at long last. It is a fact that the present search for such a theory in one way or another employs quantum methods which liberally apply probabilistic methods.

Come what may, if we define chance as an event to whose occurrence we assign the probability of less than one (but not equal to one, cf. section 7.2), its existence in the mathematical structure of the world is inevitable. Even the macroscopic world may not do without it.

4. The structure of chance

In this way, the intuitive notion of chance joins the network of mathematical structures. After all, probability is a mathematical structure and, moreover, a very elegant one (part of measure theory). Whenever we apply probability to the macroscopic world, we naturally have in mind the classical probability calculus formalised by Kolmogorov. At this level of reflection, we may, strictly speaking, eliminate the concept of chance from our considerations. It is enough to take into account events to which we assign an appropriate probabilistic measure (less than one). Thus, a certain indeterminacy or fuzziness linked with the notion of chance disappears. But the world and situations to which we apply probability calculus are not purely formal structures and the reference of probability theory to the real world opens up the circle of issues connected with the interpretation of probability. This does not prevent us from maintaining that random events (in the sense discussed above) constitute part and parcel of the mathematical structure of the world. This mathematical structure of the world – or in brief, its mathematicality – should be understood in this context exactly in the same way as in the context of other mathematical theories. We have thus certain purely formal mathematical structures, by means of which we model certain properties or certain aspects of the world, i.e. we determine the relations ('bridging principles') between certain quantities that constitute the elements of a given mathematical structure and measurable properties of the world. If a mathematical structure under consideration belongs to a measure structure with the measure being normalised to unity, then we are talking about the probabilistic properties of the world.

The entire reasoning presented above is supposed to demonstrate that chance (an event whose probability of occurrence is less than one) does not destroy the mathematicality of the world, i.e. it does not constitute a breakdown of rationality. On the contrary: it constitutes an element which is very intricately interwoven into the mathematical structure of the world. Rationality does not founder on random events. A rationality that would break down on anything would not deserve to be called a rationality.

Such a structure for chance has its own further-reaching consequences. If chance is understood in terms of probability theory, and probability theory – just as almost all mathematical theories – is subject to further generalisations (cf. Chapter 8), we must also be ready for a further generalisation of the notion of chance. Such situation already occurs in quantum mechanics. Almost from the very beginning of its existence, it was known that probabilities (less than one) appeared in it as a matter of course, but much more time had to elapse before it was understood that probability in quantum mechanics significantly differs from classical probability.

Characteristically, chance is seldom mentioned in quantum mechanics (even in philosophical and popular discourse). In my opinion, there are at least two reasons behind it. First, random events, understood as those whose probability is less than one, are nothing exceptional in it. Aristotle's commonsensical observation (cf. Subsection 1.2) that chance events are those events that happen "neither always, nor for the most part." Drawing on this intuition, chance events in quantum mechanics do not deserve to be called thus, since they occur 'almost always.' Second, probability distributions in quantum mechanics evolve in a determinist manner: if we know the probability distribution at a given moment, using the Schrödinger equation we can calculate the probability distribution at any other moment (provided that the act of measurement does not occur in the meantime). Such a probabilistic regularity does not seem to agree with the intuition of chance, so it stands to reason that quantum mechanics should refrain from using the language of chance, although, obviously we should be fully aware that it is a profoundly probabilistic theory. And it is only a matter of convention whether we decide to talk about chance or not.

Nowadays, it is beyond dispute that probability theory is inevitable whenever we try to recreate the structure of the Universe. However, the problem remains: which probability theory? There is no single, absolute probability theory. It shares its fate with other mathematical theories – it can be generalised in different ways and thereby contribute to the development of mathematics. There is the classical probability theory (formalised by Kolmogorov), there is the quantum probability theory and there are noncommutative probability theories (Chapter 8). Which ones should be included in our considerations concerning the role of chance in the structure of the Universe? We live in a macroscopic world and our daily experience has strong ties with this world, hence it seems that classical probability calculus should constitute our point of departure. Indeed, a point of departure for our investigations, but it does not necessarily play a fundamental part in the cosmic structure. All our previous knowledge of physics appears to prompt us that the basics should be sought at Planck's level (not without reason called the fundamental level) for there are small chances that any future theory of this level would vindicate classical determinism with any of its classical interpretations. The present attempts to develop a theory of the fundamental level suggest something opposite – namely that the 'cracking' of this level will be tantamount to yet another huge conceptual revolution and that if we have yet to fully succeed in our attempts, it is because they have not been radical enough. Likewise, everything seems to indicate that generalisations of the classical probability calculus will constitute a significant part of these revolutionary changes.

PART THREE

THE CODA

In musical genres, a coda constitutes the end of a motif but not necessarily the entire piece. Sometimes it serves as preparation for a counterpoint yet to follow. Something of this kind must appear in our considerations: the end of a motif, but not of the whole piece, an indication of what is yet to come, but still remains unknown. The technique of counterpoint consists in the writing of several independent melodic lines in order to bring them to a meeting point that will set off a new plot...

The present, third part of the book, has a special leading motif. It may be encapsulated in the form of a question: 'God or pure chance?' Thus we have something akin to a 'natural theology of chance.' Natural, because it is part of philosophy, and yet a theology, although without invoking the Revelation.

In order to understand the origin of contemporary interest in this problem and to steer the discussion, we will begin with a brief history of American religious fundamentalism (Chapter 10). Two completely correct terms from a theological point of view – 'creation science' and 'Intelligent Design' – have been appropriated by this fundamentalism and acquired a new, strictly parochial meaning (i.e. associated exclusively with this movement).

The question: 'God or pure chance?' clearly resonates with a Manichean echo (a heresy from the first centuries of Christianity): God does not reign over everything, there is a force which opposes Him. According to the Manichean worldview, matter is the source of evil, or in its today's version, the source of all evil is chance (Chapter 11). Over the ages, the Christian doctrine of Creation struggled against the incessantly recurring 'Manichean temptation.'

It is fairly surprising that St Thomas Aquinas anticipated as if the accusations of supporters of Intelligent Design and he devoted the entire Chapter 74 of his *Summa contra gentiles* to the thesis that "Divine Providence does not exclude chance and random events (occurring *ex casu et fortuna*)." It is worth reading this chapter carefully (Chapter 12).

What then does the role of chance in the structure of the world created by God look like? (Chapter 13). Does a small probability of occurrence of an event force us to attribute it to the direct intervention of the Creator, as was done by the supporters of physico-theology in the 17th and 18th centuries? Should the very special initial conditions indispensable for life coming into being in the Universe (anthropic principles) be neutralised on the assumption that there exists an infinite number of universes with all possible initial conditions? Does such a doctrine not oppose the idea of Divine creation? But is it still a scientific doctrine?

Next comes the time to compose an all-embracing panorama using fragments scattered throughout the book (Chapter 14). The image of the world outlined by the achievements of contemporary science is one of an evolving world. Biological evolution constitutes only a single strand in this cosmic epos. The emergence of chemical elements, the basic building blocks of life, reaches back to the first minutes after the Big Bang and to the processes that occur within the stars. The birth of planets, of which at least one constitutes the habitat of life, is a natural consequence of processes which have led from quantum fluctuations of primordial plasma to the mechanisms behind the development of galaxies and their clusters until the birth and death of stars. But evolution also consists in non-linear dynamic processes which – thanks to the sensitivity to small, random fluctuations – may lead to the growth of complexity and produce authentic novelties. The cosmic evolution is a series of necessities encoded in the laws of nature and random events indispensable for these laws to operate and to drive creative processes.

How to place the scientific vision of an evolving Universe within the doctrine of creation of the Universe by God (Chapter 15)? The term 'Intelligent Design' would solve the problem beautifully, but it has been compromised by American fundamentalists. For this reason, I prefer to use the term borrowed from Einstein – the Mind of God. God thinks mathematically and our mathematicised sciences are nothing else but an attempt to decipher this Mind. Random events are not breakdowns or symptomatic of damage to the Mind, on the contrary, they constitute the crucial points of its architecture.

Chapter 10

A short history of anti-evolutionary fundamentalism

1. The beginnings of fundamentalism

The beginnings of American religious fundamentalism date back to Ellen White (1827–1915), a visionary and founder of a religious group called the Seventh-Day Adventists. In 1864, five years after the publication of Darwin's *On the Origin of Species*, in the description of her own visions she confessed that she had witnessed the creation of the world that lasted exactly one week, the same as all the others. In early 20th century, an Adventist and a self-taught geologist George McCready Price introduced Ellen White's vision concerning the deluge in a quasi-scientific form, which, over time, also came to be widely discussed. At first, however, these views did not have a greater impact on America. Even among the conservatively-minded Protestant groups, the publication of Darwin's theory did not provoke sharper reactions. The situation changed due to a certain historical coincidence.

In 1835–1836, in Germany, David Friedrich Strauss published a two-volume book entitled *Das Leben Jesu, kritisch bearbeitet* that set about demystifying Jesus' history as described in Gospels. When the book reached the United States, it caused uproar. On one hand, Strauss' followers promoted the somewhat naively 'demythologised' religion, while on the other hand 'critical theology' was considered to be a deadly threat to religion. No wonder that it met with a strong wave of determined reaction.

In 1878–1897, in order to respond to such danger, Protestant churches in the United States organised a cycle of conferences under the joint name The Niagara Bible Conference whose aim was to define the 'fundamental truths' for Christianity. To that end, an extensive publication was brought out (12 volumes) entitled *The Fundamentals* (financed by brothers Lyman and Milton Stewart). Its main objective was to fight 'modernism,' i.e. the naturalistic interpretations rooted in German Biblical criticism. Writers contributing to the series did not have uniform views on the theory of evolution: besides decidedly critical comments, there were also milder opinions and even those favouring evolutionary ideas. Adventists were not even invited to participate in the project, yet their strong anti-evolutionary views gradually became more widely adopted in America. Around 1920, with reference to *The Fundamentals* as a project, religious-conservative views began to be called fundamentalism in America. The so-called scientific creationism (cf. below) became a part of it.

The use of Darwin's theory to justify various 'eugenic conceptions' contributed to the consolidation of anti-evolutionary views in the United States. For example, when the Virginia State Court ordered the sterilisation of Carrie Buck against her will due to her feeblemindedness in 1927, it was acknowledged to have taken place in the name of the evolutionary strategy of natural selection. Many people saw evolutionary ideas as the root cause of the revival of the German militarism, which culminated in the outbreak of World War I.

2. Monkey trials

Soon in several states, including Tennessee, local parliaments issued bans on the teaching in schools of anything "that would contravene the history of God's creation of man as taught by the Bible." In 1925, in Dayton (Tennessee) began the famous 'monkey trial' that grew into a symbol of the fight of the forces of progress against religious obscurantism. Teacher John Scopes was accused of breaking the law that forbade the teaching of evolution at schools. When his defence counsel Clarence Darrow called upon experts during the trial who were to certify that the theory of evolution was a scientific one, the court had the argument dismissed on the grounds that the subject matter of the trial was not whether or not the theory of evolution was true, but whether the law had been broken. Then Darrow resorted to ridiculing the Tennessee state counsel William Bryan, who was easily shown to be ignorant on basic Biblical issues. The court sentenced Scopes to a $100 fine. A week after the verdict William Bryan died.

Darrow was satisfied with the court ruling since he intended to appeal to the Supreme Court in order to gain publicity for the case. In the meantime, the Tennessee appeals court found procedural faults with the trial and revoked the verdict thus making the appeal impossible.

Several other states followed suit. The theory of evolution practically disappeared from school books. American society, especially its lower classes, supported these tendencies. The situation did not change until around 1960. First of all, the theory of evolution became more firmly established. It became clear that modern biology was unthinkable without this theory. The rapidly developing study of genetics delivered new arguments and, together with the theory of evolution, shaped the core of sciences about life. In the United States, perhaps to an extent greater than in other countries, the situation in science depended on politics. In the 1960s, the Cold-War arms race and the domination of the Soviet Union in space exploration made the authorities realise the magnitude of the problem of the poor quality of American education. The issue of teaching evolution was again back on the agenda.

This time, the conflict began at the other end. The legal interpretation of the First Amendment to the American Constitution provides for a strict neutrality of school instruction with respect to all religions. Since the bans on the teaching of the theory of evolution stemmed from religious motives, they were to be deemed unconstitutional. The debate on the subject grew heated when, in 1968, the biology teacher Susan Epperson sued the state of Arkansas for, among other things, the fact that its legislation forbidding her to teach the theory of evolution violated her right to the freedom of expression. Another spectacular trial began, which was promptly dubbed 'Scopes II' by the media.

However, the lawsuit against the state of Arkansas differed very much from the lawsuit against John Scopes. Now, the opponents of the theory of evolution had to prove that their version of how life began was not a religious doctrine. They adopted a strategy developed by the Yale University law student Wendell Bird. In 1978, he wrote an article in the "Yale Law Review" in which he demanded equal treatment (i.e. 'equal time' in school curricula) for the theory of evolution and the creation science (as he called the literally understood Biblical version of the origin of life). In turn, the American Civil Liberties Union (ACLU), which had already been involved in defending science against the anti-evolutionists since the times of the Tennessee lawsuit, hired the best New York defence counsel and numerous outstanding experts who were to testify.

A bizarre outcome followed: the court was supposed to decide which of the two conceptions (maybe both of them?) was a scientific theory. As we know, in the philosophy of science there are ceaseless debates discussions concerning the criteria of scientificity, and now the decision was in the hands of a judge (one may wonder if he was aware of these controversies...). Witnesses for the prosecution – that is, those defending the theory of evolution – included Catholic bishops and theologians, Presbyterians, Methodists and others. The lawsuit abounded in episodes which provided ample fodder for the sensation-seeking press. At a certain point of his testimony, the well-known theologian Langdon Gilkey, professor at the University of Chicago, declared that from the point of view of Christian theology, Act 590 that forbade the teaching of the theory of evolution at schools in Arkansas, was a heresy, since it treated creation science as a secular science, whereas it is was not known whether the Being that had created the world and living organisms out of nothing was a God of the Judeo-Christian tradition or a blind and ruthless force as preached Gnostics and Marcion's heresy in early Christianity. The defence counsel predictably responded that creation science was a secular science and not a religious one, and whether or not it was a heresy was immaterial for the proceedings at hand.

The theory of evolution was defended by well-known scientists, including Jay Gould, Francisco Ayala and Michael Ruse. Witnesses for the defence included only one well-known scientist named Chandra Wickramasinghe. He spoke against Darwin's theory, because he maintained that life was a common phenomenon in the Cosmos and it had been from the Cosmos that Earth "got infected" with life. But he had nothing to say in defence of creation science. He may have been called as a witness by opponents of the evolution theory only because they had thought that every opponent of Darwin must be in favour of creation science. Ayala thought it fit to instruct attorney David Williams in this matter:

> Mr dear *young man* – Ayala rebuked attorney *David Williams* – negative criticisms of evolutionary theory, even if they carried some weight, are utterly irrelevant to the question of the validity or legitimacy of creation science. Surely, you realise that *not* being Mr Williams in no way entails *being* Mr Ayala.[147]

Judge William Overton in his verdict not only revoked Act 590, but also ruled that creation science was not a science.

In the meantime, the campaign moved to Louisiana. This time Wendell Bird was also the lawyer who became the main driving force behind the cause of creation science. In Louisiana, the equivalent of Act 590 was called the Balanced Treatment Act. It provided for a 'balanced treatment' of evolution and creation science and both doctrines were supposed to be presented "as theories and not as proven scientific facts." Bird attempted to file a spectacular lawsuit. Al-

---

[147] Quoted from K.W. Gibberson, *Saving Darwin*, Harper One, New York 2008, p. 102.

though formally the lawsuit did not proceed, the decisions of judges were unfavourable for Bird. The disappointed lawyer submitted his case to US Supreme Court.

The trial took place in December 1987. Bird called upon several experts coming mainly from the local colleges. The ACLU mustered a large body of first-class scientists. The National Academy of Sciences and seventeen state academies sent their letters of support. A document signed by seventy-two Nobel Prize laureates was also submitted. Despite Bird's huge eloquence and legal erudition, he lost. The Supreme Court decided that the law of the state of Louisiana ordering "equal treatment" of creationism and the theory of evolution was unconstitutional, harmful for education and itself constituted a religious doctrine. Of the nine judges, two submitted votes of dissent. From the legal point of view, the matter was closed in principle. At least in principle, because life tends to be more complex than regulations.

3. Intelligent Design

If anti-evolutionism was to gain admission to American public schools, inevitably it had to appear in the guise of something that could not be identified with any religious doctrine. The version concealed under the name of creation science turned out to be unsuccessful. A fierce promoter of the new movement was Philip Johnson, professor of law at the prestigious School of Law Boalt Hall at University of California, Berkeley, and the movement adopted the term Intelligent Design as its slogan.

Johnson changed his tactics. He directed his attacks not so much at the theory of evolution as at scientific naturalism. As long as naturalism constituted a methodological foundation of science, arguments invoking God stood no chance. But in his fight against naturalism he did not invoke religious arguments, but complex phenomena in nature which – in his opinion – could not be explained by natural reasons (e.g. by means of the theory of evolution). They were, therefore, traces of an Intelligent Design concealed at the basic level of nature's operation.

This time, a cluster of intellectuals gathered around Johnson, including, among others, the mathematician and philosopher William Dembski, the biochemist Michael Behe, biologists Dean Kenyon and Jonathan Wells. In Seattle, the Discovery Institute was established that still operates as a 'scientific powerbase' to the entire movement.

Another problem arose in November 2004, when local school district authorities in Dover, Pennsylvania, imposed an obligation on teachers to read a special declaration devoted to Darwin's theory at the beginning of the biology course (in the 9th grade of the American system). The declaration stated that Darwinism was a theory, not a fact, and that it contained a number of gaps that could not be explained by the theory. Another explanation for the beginnings of life was Intelligent Design. Those interested in this conception could obtain more information on the subject from the book entitled *Of Pandas and People*.

ACLU challenged this instruction in a court of law. Since creation science had already been forbidden by the law, the matter to be decided now was whether ID was a masked form of this ideology and whether or not the book *Of Pandas and People* advanced creationism. This time again the course of the trial was extremely unfavourable for the defenders of ID. During the hearings it turned out that members of the state board of education included numerous supporters of creation science. William Buckinghamshire, chairman of the committee responsible for school curricula, denied the allegation that he was a believer in creationism, but numerous witnesses and articles from the daily press testified to something exactly opposite. And when he declared that he did not know where the funds for the purchase of *Of Pandas and People* came from, it turned out that he himself collected money for this purpose in his church. The book was portrayed as a manual of Intelligent Design not as propaganda of creation science in disguise, but it was soon proved that the first editions of the book had argued directly in favour of creation science. In subsequent editions, the old terms were simply replaced by new ones (corresponding to ID), with the rest remaining unchanged.

The trial was so unfavourable for ID that the Seattle Discovery Institute distanced itself from the entire matter.

In his verdict, Judge Jones underscored the fact that ID was nothing other than a continuation of the strategy pursued by the supporters of previous forms of creationism.

4. God or pure chance?

The supporters of creation science quite often contrast God with 'pure chance'. Was the Universe created by God or did it emerge on the strength of pure chance? – such a question pops up in different shapes on numerous websites if we type in an appropriate sequence of words in a search engine. Most of the time, this question rings with the undertones of a rhetorical question, because how could such a beautiful and complex Universe have emerged by way of chance? The strategy of supporters of Intelligent Design has changed: they no longer preach directly that chance contravenes God, but that chance cannot explain all the complexity characteristic of the Universe. The 'irreducible complexity' (a term used by supporters of ID) constitutes the trace left by Intelligent Designer in the world.

As we have seen, the target of attacks by ID promoters is the principle of naturalism adopted by official science. If naturalism – i.e. the postulate that everything should be explained exclusively in terms of natural causes – is established as the main methodological principle of science, then God (the Intelligent Designer) is eliminated not on the strength of any arguments, but on the strength of a certain premise, yet such an approach is unscientific.

Indeed, it is true that one of the main methodological principles of science is naturalism, but –as Karl W. Giberson[148] rightly noted – it does not constitute an a priori assumption resembling axioms in a formalised system. Naturalism has taken root, because it turned out to be exceptionally effective in the history of science, hence for clearly pragmatic reasons. Moreover, whenever science departed from naturalism, it invariably turned out afterwards that it had led to serious errors.

What remains is the matter of terminology, but an important one, because its oversight leads to consequential misunderstandings. This sad story of the fight of American religious fundamentalism bears witness to two serious appropriations.

In Christian theology, the term creationism (derived from Latin *creatio*) has been a technical one since the early patristic epoch. One of the time-honoured branches of theology is represented by the treatise *De creatione* (also often called *De Deo Creatore* – Of God, the Creator). Every student of theology knew what creationism was and had no negative associations with it. At present, in popular discourse creationism is opposed to evolutionism and is treated as a synonym to backwardness and being anti-scientific. It constitutes an obvious consequence of the fundamentalist attempt to transform creationism into a pseudo-scientific conception called creation science. Thus, the noble term *creatio* was appropriated by religious fundamentalists and today it is next to impossible to use it in an ordinary theological context without supplementary explanations and provisions.

The second term thus appropriated was Intelligent Design. Naturally, every person who believes in God is aware of the fact that God, in creating the world, had an 'intelligent design,' although it certainly should not be understood in the same way as e.g. the structural design of a jumbo jet. What comes to mind is Einstein's *Mind of God* and his statement that the purpose of science is, in

---

[148] *Ibid*, pp. 159–160.

fact, nothing but to decipher this Intention. Today, we cannot use the term Intelligent Design any more with reference to the creation of the world without conjuring up immediate associations with the ideology advanced by Johnson and the Discovery Institute. Their conception of 'intelligent design' stands in a radical contrast with what Einstein intended to convey in his own concept of the Mind of God. The Einsteinian Intention is a Perfect Intention, without gaps or holes that would have to be supplemented by special interventions. On the contrary, *Intelligent Design* assumes that whatever cannot be explained by science – any gaps in our understanding of the world – must be a trace of Intelligent Design. Well, not really that intelligent, since it has gaps...

In further chapters, I wish to reflect on how chance and random phenomena are integrated – without gaps or holes – into the Mind of God embodied in the Universe in which we live.

Chapter 11

The Temptation of Manichaeism

1. Previous disputes in new guises

Does God command everything? Is there anything in the work of creation that God did not succeed in doing or anything that opposes Him? Or perhaps the world, at least partly, is not the work of a Good Creator, but of some Force of Evil or an Anti-God? At the source of these questions lies the human experience of evil, and in the early period of Christianity they appeared so often that contemporary theological reflection over the truth about creation supported by dogmatic definitions of the young Church focussed on underscoring the fact that God had created EVERYTHING and that all the work of creation was GOOD. This doctrine was directed against the then rapidly spreading heresy of Manichaeism, which drew its vital juices from multi-layered and influential – especially in the East – current of thought called Gnosticism.

In late antiquity, Manichaeism died out in the Christian West, but every so often it has enjoyed revivals. For example, the doctrine of the medieval sects of the Cathars and Albigensians carried distinct traces of Manichean views. The temptation of Manichaeism in its ancient form returns now and again. Sometimes it does so in disguise, maybe even those who profess it are unaware, but distinct enough that it is hard not to notice. If the supporters of creation science posit an exclusive disjunction: either the Universe and life results from blind chance or they are the work of God, they implicitly assume that blind chance is a force opposed to God (and its opposition can be quite effective), something akin to an anti-god. The fact that they are in favour of God and against chance in no way changes the tacitly adopted assumption that God has a rival and that He does not command chance.

The doctrine of Intelligent Design finds itself in no better situation on that score. Where supporters of creation science would like to see the operation of chance, ID supporters detect an 'irreducible complexity' that cannot be explained by the recourse to natural causes (in a naturalistic way), which is thus a trace of the Intelligent Designer. But, in fact, this comes down to a strategy applied in creation science, since we are faced with an exclusive disjunction: either 'irreducible complexity' should be explained with recourse to the operation of chance (this element of the disjunction is not acknowledged by ID supporters), or as a trace of Intelligent Design. The entire reasoning clearly betrays the spectre of Manichaeism.

The ID programme officially announced that it distanced itself from religious motivations, but – as we saw in the previous chapter – a distinct motive for the emergence of creation science was the wish to defend Christianity against the threats posed by atheistic science. It sounds ironic that the fundamentalist defence of Christianity had to appeal to elements of a doctrine which was considered to be a heresy in the first ages of Christianity.

2. Gnostics or 'those who know'

The word 'gnostic' derives from the Greek *gnosis* meaning cognition or knowledge. Gnosticism, developing rapidly in the first ages of Christianity, was a collection of fairly diverse views. It was united by the faith – that had many factions and variants – that salvation can be attained on the strength of 'poured knowledge' concerning the secret of the Universe and various magic formulas and rites ensuing from such knowledge. There are people 'who know,' and others who 'do not know,' and 'those who knew' constituted a privileged class. Gnosticism borrowed many aspects from different religions existing at the time, especially from Christianity, but it distorted them in their own way. Everything was strongly tinged with pantheism, but with a pantheism of a specific kind. The world had divine features, but of a god who was depraved in a sense and was unable to control the negative features of matter.

The origins of Gnosticism go back to the Far East and vanish in the twilight of history. It met with a favourable response in the Persian kingdom and the Roman Empire. It also constituted a serious threat to early Christianity.

A substantially more organised form of Gnosticism was Manichaeism. This religion developed intensely in 3rd–4th centuries A.D.. Its founder was one Mani (Greek *Manys*, Latin *Manes* or *Manichaeus*). It was not his proper name (which remains unknown), but a kind of prophet title, which he probably gave to himself. He was born in 215 or 216 in Babylonia where he also spread his teachings. With varying degrees of success, he preached his religion at the royal court and, eventually, he managed to win favour with Ahura Mazda I, whose reign, however, was short, and his successor Bahram I threw Mani into prison, where the self-proclaimed prophet died in 276 or 277 awaiting execution.

One of the characteristic features of Gnosticism was dualism, i.e. the belief in the existence of two opposing principles: the principle of good and the principle of evil. This belief adopted different forms – starting with a radical dualism postulating the existence of a god of good and a god of evil, to a milder version, in which the god of evil was weak or in a certain sense subordinated to the god of good, up to a, so to say, diluted form, in which one God existed, but somehow limited by the elements of evil. Manichaeism preached radical dualism, but clothed it in ornate and extremely mythological forms.

According to the Manichean doctrine, before the heavens and the earth came into being, two principles existed: the Principle of Good and the Principle of Evil. The Principle of Good – the Father of Light inhabits the Area of Light that stretches into infinity, but is on one side limited by the border with the Kingdom of Darkness, which also stretches into infinity in the remaining directions. The King of Darkness is never called

God, but is endowed symmetrically with all the features of God of Light – with the opposite sign. The description of both principles is extraordinarily extensive, mixing in elements borrowed from various religions: Zoroastrianism, Buddhism, the Old Testament and Christianity.

Two powers, Light and Darkness, would have been able to coexist peaceably forever, had the King of Darkness not invaded the Kingdom of Light. This caused a war, a struggle between the forces of light and the forces of darkness. During the fight, the King of Light gave birth (by way of emanation) to the Mother of Life, who, in turn, gave birth (by emanation) to the Original Man. The description of the struggle between Light and Darkness is unusually detailed and mythologically colourful. Throughout the struggle, new worlds come into being. They are shaped by the Mother of Life from the corpses of the Sons of Darkness killed in the battle. Etc., etc. The next stages of the fight represent a metaphorical explanation of the structure of the planetary system. The origin of life and the first humans, Adam and Eve, are interwoven into a succession of obscene and cruel details. Adam's father is a masculine devil, but Adam's body contains numerous "germs of light." In order to help man, the King of Light sends the Saviour, Jesus. He is the personification of the cosmic light diffused throughout the world, which is born, suffers and dies every day.

The salvation of people takes place through gnosis. Only 'those who know' will share in the Kingdom of Light.

3. Mathematics against gnosis

According to the Manichean doctrine, 'those who know' are divided into two classes: 'the chosen ones' are those who, to a substantial extent, have already managed to extricate themselves from the entanglements of the material world. They neither marry nor participate in public life. On the other hand, 'the Listeners' only partly follow Mani's precepts, hoping that in their future incarnations they will rank amongst 'the chosen ones.' The young Augustine was a follower of the Manicheans.

Without question, St Augustine belongs to the most outstanding thinkers in the history of philosophy. The fact – which we may find surprising today – that as a young man he succumbed to the Manichean doctrine, only proves how attractive it must have been at the time. I think that the key to its understanding is the problem of evil. Is it not the case that many our contemporaries who reject the existence of God (or they do not accept Him), because they cannot reconcile His existence with the existence of suffering and evil, do so for reasons not so very different from those of the Manicheans?

But the Manichean mythology was no longer enough for Augustine. In his search, he referred to Platonic and Neoplatonic texts.

> And being admonished by these books – he will write afterwards in his *Confessions*[149] – to return to myself, I entered into my inward soul.

By the same token, Augustine's interest in natural sciences undermined his confidence in the Manicheans. It is not known how deeply Augustine penetrated the contemporary astronomy and other Greek sciences, but without doubt he was interested in them and knew quite a lot about them. In Book V of his *Confessions* he wrote:

> Yet I remembered many a true saying of the philosophers about the creation, and I saw the confirmation of their calculations in the orderly sequence of seasons and in the visible evidence of the stars. And I compared this with the doctrines of Mani, who in his voluminous folly wrote many books on these subjects. But I could not discover there any account, of either the solstices or the equinoxes, or the eclipses of the sun and moon, or anything of the sort that I had learned in the books of secular philosophy. But still I was ordered to believe, even where the ideas

---

[149] Book VII, Chapter 10, 16. All the English translations of the quotations by A. C. Outler, *Augustine: Confessions*, 1955 (accessed at: http://www9.georgetown.edu/faculty/jod/augustine/conf.pdf).

did not correspond with – even when they contradicted – the rational theories established by mathematics and my own eyes, but were very different. (V.3.6).

The breach was made.

Thus the zeal with which I had plunged into the Manichean system was
checked, and I despaired even more of their other teachers, because Faus-
tus who was so famous among them had turned out so poorly in the various matters that puzzled me. (V.7.13)

The historian Olaf Pedersen noted that:

It is difficult to say how well versed St. Augustine was in mathematical
astronomy, but the fact that it agreed with observations certainly helped
him to leave the Manichean community and join the orthodox faith; perhaps this was one of the most important services rendered by Greek science to the ancient Church.[150]

Augustine's respect for science remained with him forever. When he
was writing his *De Genesi ad Litteram* directed against Manichean mis-
representations of the understanding of the Biblical story about Creation,
he formulated the following principle:

When an apparent conflict arises between a strongly supported scientific
theory and some item of Christian doctrine, the Christian ought to look
very carefully to the credentials of the doctrine. It may well be that when
he does so, the scientific understanding will enable the doctrine to be reformulated in a more adequate way.[151]

---

[150] O. Pedersen, *The Two Books, op. cit.*, p. 94.
[151] E. McMullin, *Introduction: Evolution and Creation*, in *Evolution and Creation*, ed.
by E. McMullin, University of Notre Dame Press, Notre Dame, IN 1985, p. 2. This
principle was formulated by Augustine in *De Genesi ad Litteram*, Book I, Chapter 21.

Augustine's motivation can be understood. God is the author of both the Bible and the *Book Nature*, so there can be no true conflict between the two. If we have a reliable knowledge about the world and it appears to conflict with the Biblical text, then we should seek a proper reinterpretation of the passage. The more so that in numerous places of the Bible there are obvious metaphors to be found. Augustine noted that if we did not proceed in accordance with his rule formulated, we would expose the Christian doctrine to ridicule in pagans' eyes. This may constitute an echo of Augustine's personal experience, when the doctrine of Manicheans was disgraced in his eyes when compared with the then contemporary science.

It is worth bearing in mind St Augustine's principle in today's discussions concerning the areas of conflict between science and religion.

4. Is God the Lord of all?

Manichaeism was not only a personal experience of St Augustine's, it also painfully affected the young Church. Interestingly, the first preserved verdict of the Church concerning the issue of Creation was the decree issued by the Synod of Braga directed against Priscillian's Manichean doctrine and his sect. The sect appeared around 375 in Spain under the distinct influences of Gnosticism and Manichaeism. In Spain, it spread amongst the clergy. In 563, a synod of bishops was convened in Braga (Spain) to deal with numerous matters concerning the cult and Church discipline. The canons issued by this Synod in question included several targeted at Manichean dualism preached by Priscillian. Several examples follow below:[152]

Canon 11. If anyone condemns human marriage and has a horror of the procreation of living bodies, as Manichaeus and Priscillian have said, let him be anathema.

---

[152] Translations of canons into English from the Catechism of the Catholic Church, (accessed at: http://www.catecheticsonline.com/SourcesofDogma3.php)

Canon 12. If anyone says that the formation of the human body is a creation of the devil, and says that conceptions in the wombs of mothers are formed by the works of demons, and for this reason does not believe in the resurrection of the body, just as Manichaeus and Priscillian have said, let him be anathema.

Canon 13. If anyone says that the creation of all flesh is not the work of God, but belongs to the wicked angels, just as Priscillian has said, let him be anathema.

It may seem that the doctrine of the Divine creation of the world is so dogmatically important and so imbued with theological contents that it should constitute the focus of the teaching authority of the Church, yet references to it in the official Church documents are few and far between. The few that do appear concern, in principle, only one topic. The theme touched upon by the Synod of Braga was subsequently undertaken by other synods and councils. They invariably did it as a matter of expediency, whenever 'the temptation of Manichaeism' materialised. The same doctrine was repeated and developed by the Fourth Council of the Lateran (1215) and later by the Council of Florence (1438–1445). The same was the case with the Second Vatican Council (1870), which, while stigmatising errors of the time, apart from the condemnation of materialism and various forms of emanationism, once again underscored that all beings – without exception – come from God.

Certainly, from the very beginning, the truth about the Divine creation of the world in all its theological splendour was frequently an object of reflection on the part of the Church Fathers and Christian writers,[153] but did not find its way into the Magisterium because it was not deemed necessary. In principle, the Magisterium speaks when it feels forced to defend a certain truth or to explain a certain truth against misinterpretations propagat-

---

[153] Cf. e.g. H. Pietras, *Rdz 1 u Ojców Kościoła (Gen 1 According to Church Fathers, in Polish)*, in *Początek świata – Biblia a nauka (The Beginning of the Word – Bible and Science)*, M. Heller, M. Drożdż, eds., Biblos, Tarnów 1998, pp. 83–100.

ed in a given age. It is symptomatic that, essentially, only one aspect of the Christian doctrine about the Creation was – and so often is – 'under threat.' It concerns the dogma that God is the Lord of all without any exceptions.

Let us return to the question asked by the supporters of anti-evolutionary fundamentalism: Is the Universe the work of God or of pure chance? The theory of evolution attributes too large a part to chance and therefore excludes God, in consequence, in order to defend faith in God one needs to eliminate (creation science) or at least significantly restrict (Intelligent Design) the theory of evolution, which tacitly amounts to an admission that Manichaeism (in its updated version) is, in fact, right.

Chapter 12

Ex casu et fortuna

1. Against the pagans

Two words that appear in the title of this chapter – *casus* and *fortuna* – were used by St Thomas Aquinas in order to convey what today we call chance. *Casus* in Latin means exactly chance, and among the different meanings of the word *fortuna* in the Latin dictionary there are chance, event, fate, coincidence; today we might even say 'a random event.' St Thomas did not preoccupy himself with probability calculus, but wondered about whether Divine Providence excludes the events that occur *ex causa et fortuna.* He devoted several chapters of his work *Summa contra gentiles*[154] to this issue. Since this is the issue with which we shall deal in the third part of this book, let us focus somewhat more on relevant portions of St Thomas' work. Naturally, we must be ready for the completely different context in which Thomas develops his argumentation, which is that of scholastic thought. Since the times of St Augustine, concepts had evolved, became more technical, and the style changed from a narrative interspersed with arguments, was forced into an orderly framework of the scholastic discourse.

---

[154] *Summa contra gentiles seu de veritate catholicae fidei.* Domus Editorialis Marietti, Taurini 1938. (all English quotations come from: *Summa contra Gentiles.* Translated by Anton C. Pegis et al. 1955; Ind.: U. of Notre Dame Press, rpt. Notre Dame 1975).

Summa contra gentiles is addressed to 'pagans,' i.e. those who do not acknowledge the Christian revelation, so in the discussion – Thomas notes – one ought to invoke 'natural reason,' *cui omnes assentire cogantur* ('which all are obliged to assent to,' Liber I, caput 2). How optimistic the latter part of this sentence appears.

2. How comes evil?

We will focus our special interest on Book III, Chapter 71. It is here that St Thomas poses the problem which – echoing Boethius – he formulates as a question: "If there is a God, how comes evil?" Considering this question, Thomas wants "to take away the occasion of the error" to which Manicheans succumbed in adopting two principles: good and evil, "as though evil could not have place under the providence of a good God."

As usual, St Thomas conducts all analyses within the framework of his own system. In his ontology, God is the First Cause of all, but in His own actions He does not reject the secondary (remote) causes and it may well be the case that a secondary cause will produce a defect in the result it has brought about, but this would be the result of an imperfection inherent in the secondary cause, not in the First Cause. Moreover, if all the secondary causes were the best out of all the possible ones, there would be no differences among them in degrees of goodness, and a gradation of goodness is better than a general uniformity. The perfection of the world requires thus a gradation of goodness, in consequence, certain secondary causes may be worse than others, and thus may produce 'defective results.'

Besides, the good of the whole should take precedence over the good of individual parts. "It belongs, then, to a prudent ruler to neglect some defect of goodness in the part for the increase of goodness in the whole." Just as a composer might do, who inserts a moment of silence into a cantilena in order to enhance the whole. What also needs to be considered is that there are many good things in the world that would not be possible unless there were some evil things, e.g. patience in the face of persecution.

Nevertheless, is God not responsible for the evil that is caused by secondary causes? After all, He is their First Cause. Yes – replies Thomas – all causes derive their existence and "the power of causation" from God, but defects in their operation originate exclusively from them.

Let us take a closer look at the style of Thomas' reasoning, which is very scholastic. First of all, as I have already mentioned above, he is systemic through and through, in consequence, for someone who does not accept the whole system, the reasoning has no "captivating power." In those days, the system was identified with "knowledge as such" (so the system was absolutised). It was not until considerably later that the achievements of the methodology of science revealed significant limitations inherent in the systemic knowledge. Moreover, even within the system, a conceptual analysis and logical reconstruction of individual reasonings would be in a position to reveal the flaws of the latter.

As far as the contents of Thomas' reasoning is concerned, it is only a single small step removed from Leibniz's argumentation, who maintained that God had created the best of all possible worlds. Indeed, Thomas' reasoning shows that the Lord decided to allow for the possibility of a 'defective' operation of secondary causes in order to maximise the extent of goodness in the world, or – using Leibniz's language – out of all the possible worlds, He decided to choose 'for implementation' the best one – yet not the best one in an absolute sense, but relative to the harmonisation and conditions that applied to all its constituent elements.

The question borrowed from Boethius, with which we began reading this chapter of the *Summa contra gentiles* (and which is actually found at the end of the chapter), introduces a touch of dramatic tension to considerations otherwise dry and devoid of emotion. "If there is a God, how comes evil?" In his reply, Thomas turns the reasoning the other way: "If there is evil, there is a God," because if there was no God, there would be no difference between good and evil. A response indeed worthy of Dostoyevsky and his well-known statement: "If God does not exist, everything is permitted."

3. Contingency and the free will

The two following chapters of the summa *Contra gentiles* are devoted to the discussion of the relation of contingency and necessity of the world (Chapter 72) and the free will (Chapter 73) to Divine Providence. The concept of contingency is of key import for Thomistic ontology. The world is contingent, i.e. it does not exist by necessity: its existence and the way it is constitute the result of the Creator's free decision. This ontological thesis has important consequences for the system according to which the world operates. The secondary causes may act in a necessary manner (due to the necessity imposed on them by God) where an effect always succeeds a given cause; or in a mechanical manner: when something can hinder the occurrence of an effect. In principle, the same arguments quoted by Thomas in connection with the problem of evil apply to this situation as well. God's strategy of acting through secondary causes does not exclude the contingent operation of secondary causes. Moreover, according to Thomas

> It would be incompatible, then, with divine providence to which the establishment and preservation of order in things belongs, if all things came about as a result of necessity. (72.4)

If such an absence of necessity appears in the inanimate world and in the world of creatures not endowed with reason, all the more it has its place in the world of man endowed with free will. Free will introduces to the 'order of the world' a certain element of indeterminacy and thus may upset this order. Here, Thomas' arguments invoking the operation of secondary causes still apply with the provision that now the secondary cause (man) – apart from his capacity to act improperly (i.e. producing defects) – acquires an authentic autonomy, because he may choose to act freely. Although freedom of the will is associated with a genuine risk of infringement and even of partial destruction of the 'order of the world,' it is a great gift from Providence.

4. A meeting in the market

It is only in this broader context that Thomas places the problem of random and chance events. In keeping with his style adopted in the *Summa contra gentiles*, the title of each chapter always contains a statement that Thomas defends in it. Chapter 74 is titled *Quod Divina Providentia non excludit casum et fortunam* (That Divine Providence does not exclude Fortune and Chance).

By random and chance events – following in Aristotle's footsteps (although he does not invoke Aristotle directly) – Thomas understands such events that "do not happen always, perforce, or the most often" (cf. Chapter 1.2). That Providence allows such events, follows even from Chapter 72, in which Thomas argues that Providence does not exclude events that do not occur by necessity. Apparently, however, Thomas considered this issue to be important enough to merit his extra attention.

Since causal chains may intersect, they form a distinctive network. It may happen that at an intersections of such chains, two (or more) causes meet and one may hinder the operation of another. In such a case, we say that a chance event has occurred. Here, Thomas (almost literally) quotes the example given by Aristotle (Physics II, 196a, cf. Chapter 1.2) about a man who goes to the market in order to buy something, and meets his creditor who came to the market with a different purpose. Thus we have two causal chains: two individuals coming to the market, both for their respective strictly determined reasons, but their meeting is accidental. Chance arose from the intersection of two causal chains.

Moreover, the sets of causes and results are arranged hierarchically, which may also lead to random events. They occur when a higher-order cause modifies the operation of the cause or causes of the lower order. In the Thomistic ontology, an important part is played by the distinction – adopted from Aristotle – between the substance and accident (*accidens*). Speaking very generally, substance is all that belongs to the essence of a certain being, whereas accidents are properties of a certain being that do not belong to its essence. For example, the conditions of being an animal and possessing rationality belong to the substance of man, but man's skin colour is an accident. This distinction entails the distinction of causal operations. A certain cause may produce substantial or accidental effects (*per accidens*). The latter mode of operation of causes constitutes the source of chance in the world. The entire structure of causality and its hierarchical arrangement is rationally intended by Divine Providence.

5. St Augustine's warning

Let us try comparing the short overview of the history of creation science and Intelligent Design made in Chapter 10 with St Thomas' struggles with the issue of Providence and chance. Let us also return to some of St Augustine's considerations. If we realise that St Augustine (the problem of Manichaeism) and St Thomas Aquinas (random events) are universally considered to be the most outstanding minds in the service of Christian theology, we can see all the more clearly that the fiercely anti-evolutionary fundamentalism contrasts with the main line of development of theology mapped out by these of two thinkers.

Let us reiterate the rule formulated by St Augustine who requires that in case of a contradiction between a Biblical statement and a well-established scientific truth, we should invoke a metaphorical interpretation of the Bible (cf. Chapter 11.3). What is interesting is the motivation behind this rule. Augustine warned that if Christians did not apply this rule in practice, they would expose their own religion to ridicule by the pagans. Anti-evolutionary fundamentalists have blatantly violated this rule and indeed, what followed was what Augustine warned Christians against – the Christian religion was derided. It is not only in the case of the 'monkey trials,' but also by today's widespread attribution of views to Christianity that manifestly contradict science. The damage caused by anti-evolutionary fundamentalism in this respect is hard to gauge.

The moral concealed in St Augustine's rule may be treated as a special case of a more general regularity. The fact that the development of science debunks various myths requires no substantiation. Religious beliefs even more than other kinds of convictions tend to become permeated by myths. Among the many reasons for this phenomenon, there is one that I would call an ontological one: since the content of religious beliefs concerns transcendental things, which are by their nature fundamentally inexpressible, yet we must somehow think and talk about them, we do it by means of mythical images while seldom realising it. The development of science corrects or outright overthrows many such mythical images, accordingly, theologians speak about the cleansing role of science. This process may sometimes be painful and may be interpreted in terms of a conflict between science and religion, but its positive impacts are hard to overestimate.

Chapter 13

The God of probabilities

1. The ontology of probability

We have already talked about the strange feature of the human mind (Chapter 7.7), which treats a high (*a priori*) probability of the occurrence of an event as the sufficient condition for its occurrence. This is a strange feature, because a purely probabilistic reasoning simulates a causal explanation. In this sense, we treat probability calculus as an ontology of sorts: the world is such that whatever is probable occurs in it. It sounds almost like a tautology since our faith in this regularity was imposed on our mind by the world. Let us remind ourselves that that the frequency stability is a property of the world thanks to which the longer the sequence of random events (e.g. a throw of a fair die), the more the frequency of occurrence of a certain event (e.g. obtaining a six) approaches a certain number (e.g. 1/6; cf. Chapter 7.7).

Moreover, it may be assumed that the mechanism that has imposed the "faith in probability" on our minds was an evolutionary one. To put it in simpler terms, the human mind, by way of interacting with the natural environment has developed a conviction that certain events occur more often than others and that it pays to use such knowledge. In this way, the structure of the human mind has become adapted to the structure of the world (i.e. to its frequency stability).

2. The problem of low probabilities

Let us open a well-known book by Roger Penrose *The Emperor's New Mind* on page 339.[155] We find there a chapter entitled *How special was the Big Bang?* In an attempt to answer this question, Penrose conducts the following reasoning. Let us imagine a huge space whose every point represents a state, in which the Universe may find itself. Such a space of all possible states of the Universe is called its phase space. Since the states of the Universe in which its entropy is small, have a small probability, their corresponding areas in the phase space are also small. Since the states of the Universe in which its entropy is large are very probable, their corresponding areas in the phase space are large. Let us now imagine that the Creator decided to create a Universe driven by 'pure chance' and to choose as the initial state of the Universe a point in the phase space indicated by the point of a pin thrown at random on the phase space. Since the areas representing a large entropy are large, the probability of the pin hitting them is high; and similarly, because areas representing a small entropy are small, the probability of the pin hitting them is low. But how low, exactly? Since the notion of entropy is a statistical one, Penrose, by invoking statistical considerations, estimated the magnitude of areas that represented such a small entropy.[156] It turns out that the volume of an area representing the small initial entropy is ten to the tenth power to the hundred and twenty third power times smaller than the volume of the entire phase space. Penrose writes:

> This now tells us how precise the Creator's aim must have been, namely to an accuracy of one part in $10^{10^{123}}$. This is an extraordinary. One could not possibly even *write the number down* in full, in the ordinary denary notation: it would be '1' followed by 10^{123} successive

---

[155] R. Penrose, *The Emperor's New Mind: Concerning Computers*, Minds and The Laws of Physics, Oxford University Press, Oxford 1989.
[156] Penrose took into account the so-called gravitational entropy, which in such cosmological considerations plays a greater role than the ordinary thermodynamic entropy.

'0's! Even if we were to write a '0' on each separate proton and each separate neutron in the entire universe – and we could throw in all the other particles for good measure – we should fall short of writing down the figure needed.[157]

In compliance with the ideology of Intelligent Design, this unbelievably low probability[158] should be interpreted as an example of an 'irreducible complexity' and considered to be a trace left for us by the Designer. Penrose is very distant from such an ideology. He treats the unusually small probability of the initial state of the Universe with a small entropy as an indication that we take advantage of in order to find the long sought-after quantum theory of gravitation. Moreover, he himself is on the track of a law that would explain this fact whose probability of occurrence is extremely low.[159]

Is it naturalism? Certainly, but it is not accepted on the basis of an *a priori* dogma, but has been developed as a methodological directive by scientific history and practice. An important warning against the breaking of the principle of methodological naturalism is the history of the so-called physico-theology. In the 17th century, at the beginning of modern science, scientists were enchanted by the wealth and complexity of phenomena shown to them by new instruments and new scientific theories. The intricate construction of an optical instrument, such as the eye of the mosquito, or specific initial conditions that must be met by the solutions to the equations that govern the motions of planets, cannot be the work of blind chance. The unusually small probability of spontaneous realisation of such configurations directed their thought to a purposeful design of the world. Such a small probability was considered to be a kind of gap in the natural course of events and the human mind – as was thought then – demanded that such a gap should be filled by 'the hypothesis of God.'

---

[157] R. Penrose, *op. cit.*, p. 344.
[158] Significantly lower than various 'incredibly low probabilities' that the proponents of ID quote as arguments in favour of 'irreducible complexities."
[159] Such a law would state that for the initial states the Weyl space-time curvature should equal zero.

Physics, which was then the most rapidly developing science and a symbol of scientific progress, became coupled with theological considerations and for this reason, this current of thought deserved to be named physico-theology.[160] But, in the long run, this phenomenon turned out to be prejudicial to natural theology and to theology in general. Rapid advances in science began to gradually fill previously existing gaps in the understanding of the world. The hypothesis of God filling these gaps became less and less necessary, until ultimately it gave way to the methodological directive that "the world should be explained in terms of the world itself." Afterwards, this rule was called the principle of methodological naturalism. It was born from the historical experience reaching further back than to the period of physico-theology. When several thinkers on the coasts of Asia Minor challenged the world at the turn of the 6th and 5th centuries B.C. by trying to understand it not by means of myths and supernatural forces, but by means of observation and the human mind, it was then that the principle of methodological naturalism for the first time started to work in the nascent science and philosophy.

Yet one should remember that the above does not constitute an ontological rule that maintains that the world is self-sufficient and needs no Creator, but a methodological principle that tells us to investigate the world using our own resources, without invoking interventions from beyond the world.[161]

3. The problem of high probabilities

There exists a certain special kind of explanation: by the reduction of chance to something that is not chance. For example, a certain phenomenon analysed in a small set may pass for something exceptional (and thus, random), but considered in a larger set may lose its ran-

---

[160] More information on the subject can be found in e.g.: O. Pedersen, *op. cit.*, pp. 258–276 ; E. Mullin, *op. cit.*, pp. 27–32.
[161] I discuss the subject of naturalism in more detail in Chapter VI of *The Sense of Life and the Sense of the Unvierse*, Copernicus Center Press, Kraków 2010, pp. 99–116.

domness. A red-haired student in a class may pass for an exception, but on the scale of the whole country it is nothing extraordinary. Translating this into the probabilistic language: the probability of there being a red-haired individual in a single class at school (in a country such as Poland) is low, but not so low if we consider the entire country. Here is where the regularity mentioned at the beginning of this chapter is in operation – a high probability simulates a causal explanation, or at least soothes our need to look for causes for events whose probability is low.

The argument based on this type of reasoning also appears in natural theology as a kind of reversal of the hypothesis of God – the Filler of gaps. Here it is. Today, we already know quite well that in order to produce a Universe similar to ours using the methods of standard cosmology, it is necessary to adopt very special initial conditions. Penrose's reasoning quoted above is only one of many similar reasonings (admittedly, an especially persuasive one). If we additionally assume that a Universe 'similar to ours' at a certain stage of its development must include a planet similar to the Earth, on which all the indispensable conditions for the initiation of a biological evolution would come true, the initial conditions of such a Universe must be yet 'more precisely fine-tuned.' Different formulations of this observation are called anthropic rules. Anthropic – because originally scientists mentioned the necessary conditions for the existence of man ('a rational observer'), although the reasoning would be no different in principle if, instead of man, we were to consider the amoeba.

Can the realisation of the initial conditions with such a low probability be attributed to pure chance? Instead, one should 'reduce' chance by extending the space in which it appears. Accordingly, one needs to consider the space of many universes, ideally, all the possible universes with all the possible combinations of initial conditions. Among them, there will certainly be found very exceptional conditions, indispensable to the emergence of life at least on a single planet. And if we make an allowance for the obvious fact that we could have come into being only in such an exceptional universe, it should be acknowledged that the problem has been solved.

In this way, the idea of the so-called multiverse was born. Originally, it was developed primarily in order to 'neutralise' the problem
of special initial conditions for the Universe,[162] and also to protect
cosmology from teleological (and theological) interpretations associated with anthropic rules. Over time, however, the idea of the multiverse took root in the consciousness of the more metaphysically
disposed cosmologists and began to appear no longer for purely ideological reasons, but as a conclusion from various cosmological speculations and models. Thus, for example, in the cosmological applications of the theory of superstrings (and its most recent incarnation,
the so-called M-theory), one of the main candidates for the theory
of the fundamental level, solutions were found which are interpreted as 'other worlds.'[163] Likewise, in the chaotic cosmology of Andrei
Linde, the descendant universes are born out of the inflationary phase
of the mother-universe.[164]

This does not mean, however, that the idea of the multiverse has
won general acceptance as a satisfactory solution to the problem of
the initial conditions. First of all, the very idea gives rise to numerous reservations and controversies. The main accusation comes down
to the doubt whether it is a scientific concept at all. Aside from the
disputes in the philosophy of science concerning the criteria of scientificity, it is universally agreed that no conception can be deemed
scientific, unless it produces conclusions that can be compared with
results of experiments or observations. Do such conclusions ensue
from the conception of the multiverse? Its proponents are trying to
persuade the others that it is indeed the case, but their arguments tend
to be limp. Other universes are – on the strength of their definition,
as a matter of fact – 'disjoint' from our Universe, i.e. we cannot have
empirical access to other universes. All the 'observational tests' pro-

---

[162] Cf. e.g. *Universe or Multiverse, op. cit.*
[163] Cf. e.g. F. Quevedo, *Lectures on String/Brane Cosmology*, "Classical and Quantum
Gravity" 2002, 19, pp. 5721–5779.
[164] Cf. e.g. A. Linde, *The Inflationary Multiverse*, in *Universe or Multiverse, op. cit.*,
pp. 127–149.

posed by defenders of the idea of the multiverse come down to determining which properties occur with the most frequency in a set of universes, and to arguing that our Universe possesses such features. Only by interpreting the criterion of empiricity extraordinarily widely (considerably more widely than it is usually done in the philosophy of science) can we consider it fulfilled with respect to the conception of the multiverse.[165] There are also many other reasons for which the notion of the multiverse can hardly be treated as sufficiently good to function properly in science.[166]

Nonetheless, the idea of the multiverse does not really have to be considered as competitive with the idea of creation of the Universe by God. From the theological point of view, God could have created both a single Universe and an infinite number of universes. Or, to put it differently: if God created one Universe, he could have created as many as He wished. There is no theological truth that would prohibit it. Moreover, Leibniz's question: "Why is there something rather than nothing?," which is usually quoted as the philosophical motivation behind the question concerning the creation of the Universe, when referred to numerous universes appears even more urgent. 'Something' about whose existence we ask questions, in the case of numerous worlds provokes even greater metaphysical anxiety than is the case with a single Universe.

4. The indifference principle

When physico-theology was born in the 17th century, it had to face an influential opponent in the form of the Cartesian vision of the world. Descartes maintained that the laws of mechanics were, in principle, sufficient to explain all the complexity of the present world on con-

---

[165] For a more extensive discussion of the issue cf. my *Ultimate Explanations of the Universe*, Springer, Berlin - Heidelberg, 2009.

[166] These reasons are very clearly discussed by George Ellis in his article: *Multiverse: Description, Uniqueness and Testing*, in *Universe or Multiverse, op. cit.*, pp. 387–409.

dition that we assume that they began to operate in a 'chaotic,' i.e. the most typical arrangement of material particles. This idea obviously ran counter to the physico-theological arguments for the existence of God.

Ernan McMullin[167] formulated the so-called indifference principle that underlies Descartes' reasoning and which even today is present in scientific explanations. This principle states that

> no constraint had to be set on the original configuration for the present-day cosmic complexity to develop from it.

Accordingly, the initial conditions leading to the present-day complexity of the world should be as typical as is only possible. Constraints mentioned in the McMullin principle would have to select from the most typical initial conditions certain more specific ones, but such constraints, as provided by the principle, should not exist. As we have seen, the proponents of the 17th century physico-theology and the present propagators of the idea of Intelligent Design try to identify such constraints and attribute them to God's creative activity. Is their argumentation justified? Let us once again invoke McMullin.

He writes that even if we were to acknowledge the portion of the argumentation advanced by the supporters of Intelligent Design that refers to nature (i.e. that there really exists an 'irreducible complexity' which cannot be explained naturally), he would still have serious reservations in accepting the Designer as an appropriate explanation for the problem. In McMullin's view, such an explanation assumes a huge ontological 'leap':

> In order to make the entire reasoning acceptable, to remove the suspicion that the proposed explanation is *ad hoc*, entirely contrived, one has to show that the hypothesis is coherent, that the entity being pro-

---

[167] *Anthropic Explanation in Cosmology*, "Faith and Philosophy" 2005, 22, pp. 601–614.

posed by way of explanation is accessible to our theoretical resources, including scientific, metaphysical and theological ones.[168]

If we are otherwise convinced about the existence of a Creator, such a conviction can be coherent with what is being done in science, but then we do not need dubious arguments that come down to the infringement of the indifference principle.

---

[168] *Ibid.*, p. 612

Chapter 14

One of the strands of cosmic evolution

1. A basket of problems which have attracted insufficient attention

Even though the supporters of Intelligent Design maintain that they fight against 'the principle of naturalism' adopted by contemporary science, in effect, the edge of their criticism is directed against the biological theory of evolution. In their opinion, it is this very theory that most acutely reveals the dilemma 'pure chance versus Intelligent Design.' In this chapter, we will resist the temptation to argue and will focus – in a positive manner – on the issue of evolution. I do not intend to present the theory even in the briefest possible outline as this would exceed both my competence and the scope of this book. However, I wish to reflect in somewhat more detail on the relationship between biological evolution and cosmological evolution, as well as on the ingrained nature of biological evolution in the laws of physics. At any rate, these three elements – the laws of physics, cosmological evolution and biological evolution – are closely interconnected. It is next to impossible to separate the laws of physics from the cosmic evolution; after all, they determine the structure of Cosmos (or constitute only its aspects), and thus its evolution as well. Besides, the Universe constitutes the natural habitat of life. Biological evolution is only a strand of the cosmic evolution. The weave of these problems immediately suggests itself, but I get the impression that it attracts considerably less attention than it deserves.

2. A sink for disorder

In physics, the state-dependent function called entropy measures the degree of disorder in a given system: the greater the entropy of a system in a given state, the greater the degree of disorder that characterises such a state. Since heat is the most unorganised form of energy, the growth of entropy is usually accompanied by the change of other forms of energy into heat. In closed systems, during irreversible processes entropy always increases until the system reaches the state of its thermal equilibrium. Entropy cannot increase any further: energy is maximally dispersed (the temperatures of the system and the environment equalise), which means that the disorder has reached its greatest possible extent. To put it more vividly, the system reaches 'the state of thermal death.'

Erwin Schrödinger in his pioneering book *What is life?*[169] noted that life feeds on negative entropy. In other words, from the point of view of physics, living organisms draw from their environment low-entropy energy mainly in the form of oxygen and food, convert it into their own well-ordered structures, and excrete high-entropy energy, i.e. waste products remaining on completion of the process of organisation, mostly as heat.

The source of low-entropy energy for organisms that inhabit the Earth are the nuclear reactions occurring in the Sun in conjunction with gravitational compression, which, by opposing the pressure of radiation, ensures that the Sun remains in a stable state. Plants on the Earth absorb a small proportion of the photons of the visible light emitted by the Sun, convert them into their own structure and emit the received energy in the form of infrared radiation (we draw low-entropy energy from the Sun through plants). But the visible light carries a considerably greater energy than infrared radiation. Since the energy balance must be preserved, the number of photons emitted by living organisms must be considerably greater than the number

---

[169] Cambridge University Press, Cambridge 1944.

of absorbed photons. Photons, once used up, are released into outer space, whose role can be compared to that of a drainpipe where the disorder generated by living organisms is pushed out. One may suspect that this drainpipe must be gigantic in order to enable the entire mechanism to operate.[170] Moreover, before the stars with properties akin to our Sun's can come into existence, the cosmic evolution must last around 10 billion years. And since the Universe keeps expanding, during all that time it will inflate to a gigantic size. Thus, in order for biological evolution to begin on at least a single planet, the Universe must be large and old.

3. The natural history of carbon

Our existence is associated with the Sun for yet another reason. Organic chemistry is the chemistry of carbon compounds. Atoms of carbon found in our bodies were produced in nuclear reactions occurring on the Sun. Without it, there would be no organic chemistry and evolution of life would have no material on which to work its wonders. What is, then, the natural history of carbon?

The nuclei of chemical elements are born inside hot stars, but observation-based research shows that stars began their evolution from a state in which their original building material – hydrogen – was 'polluted' by small quantities of light elements. This is confirmed by the theory of cosmic nucleosynthesis. Let us briefly outline its most important points.[171]

In a young and very hot Universe, quarks and gluons were not able to combine to form protons and neutrons until the temperature fell to 10^{12} K, which took place when the Universe was 10^{-4} seconds

---

[170] For a more complete description of the process, cf. R. Penrose, *op. cit.*, pp. 317–322.

[171] A short outline of the cosmic nucleosynthesis can be found in A. Liddle, *An Introduction to Modern Cosmology*, Wiley 2003, Chapter 12. A considerably more detailed discussion of the subject can be found in L. Sokołowski, *Elementy kosmologii (Elementary Issues in Cosmology, in Polish)*, ZamKor, Krakow 2005, pp. 165–203.

old. Beforehand, temperatures were so high that even if protons and neutrons had been produced, they would have been torn apart immediately. In turn, protons and neutrons were able to combine to form the nuclei of light chemical elements when the temperature fell to 10^9 K. Then the process of nucleosynthesis of the lightest elements began, which took place when the Universe was 1–3 minutes old. As a result of these processes, the following 'chemical constitution' of the young Universe was established: helium 4 constituted approximately ¼ of the total mass of the entire Universe (considering only the baryonic matter), deuterium – 10^{-4} of the total mass, helium 3 – 10^{-5} of the total mass, and lithium 7 – 10^{-10}. The balance was made up by hydrogen (whose nucleus consists of a single proton).

These theoretical calculations can be compared with the curve of abundance of chemical elements in the Universe (plotted on the basis of observations). The conformity is very good. A huge success of the theory was the conclusion that such a conformity could be obtained only when it is assumed that there are three kinds of neutrinos (electron, muon and tau). Only later was this conclusion – that there are only three kinds of neutrinos – confirmed experimentally at CERN.

So far, we have been discussing the emergence of nuclei of chemical elements. The ambient temperature was still too high for nuclei to be able to capture electrons and thus create atoms. This became possible only approximately 400,000 years after the Big Bang, when the temperature of the Universe fell to 4,000 K. It took place towards the end of the radiation era, when the Universe was filled with hot electromagnetic radiation (today, we can observe residual radiation of the kind as the so-called cosmic microwave background radiation), and around the beginning of the galactic era, when the processes leading to the emergence of stars and galaxies, began.

The process of nuclear fusion (atomic nuclei combine to form the nuclei of heavier elements) takes place in the interiors of stars. Nuclear reactions involved in this process occur in various forms, depending on the mass of a given star. The first stars contained only hydrogen and helium (and vestigial quantities of other original light elements), which gradually burnt through to form heavier elements. During the explosions of stars (the so-called supernovae), these elements were released into the interstellar medium. From the dusts thus enriched, the second-generation stars were born and the process was repeated. Two or three generations of stars are needed to produce carbon. Our Sun is one such star.

The direct building blocks of carbon are beryllium and helium. In order to produce carbon, two nuclei of beryllium and one nucleus of helium must combine. However, since beryllium is an unstable element (i.e. it decays just after it has been produced), the probability of occurrence of such a reaction is very low. Yet it increases dramatically owing to the fact that the basic state of the beryllium nucleus almost exactly matches the resonance energy present in the induced state of carbon ^{12}C. Thanks to this harmonisation, the reaction has sufficient time to occur, before beryllium decays. The process is not dissimilar to the case of a sieve which will hold water if it is poured on it quickly enough. It is a remarkable thing that the famous astronomer Fred Hoyle anticipated the existence of such a resonance in the structure of carbon, because he concluded that this was the only way for stars to synthesise carbon. After all, carbon exists, because we do. The process of combining beryllium with helium to produce carbon cannot occur inside stars such as our Sun, since the temperature in their interiors is too low. The process takes place in the interiors of stars known as red giants. That is why the Sun is a second-generation star that materialised from the ashes of a red giant.

Thus, our existence depends on stars which have evolved for billions of years, and throughout all this time galaxies move further apart. The universe must be great and old for life to have developed on even a single planet.

4. Dynamical systems

But this is not everything. The universe is governed by physics, and the relationships of biological evolution with physics are very intimate. They touch the very understanding of the notion of evolution itself. Usually, even in biology we use this notion intuitively. Obviously, having studied the biological theory of evolution in some depth, our understanding of the notion of evolution becomes clearer, acquires a richer background, but if we were to express its essence in a few words, it would be likely next to impossible.

In physics, we also come across the concept of evolution. Often we say that a certain physical system evolves or is subject to evolution, and the meaning of this in mathematics and physics is very precisely determined. The theory that deals with this is known as the theory of dynamical systems.

The essence of the evolutionary process requires that an evolving system go through successive states, but these states do not succeed one another in a haphazard manner, each state is a dynamic consequence of the previous one. A state is a collection of properties that characterise a system at a given moment. For example, if the evolving system is a moving mass point, then its state at a certain moment is described by its position and velocity at that moment. If an evolving system is a set of mass points, its state at a certain moment is characterised by the positions and velocities of all mass points at that moment. In the case of more complex systems, physics always supplies a recipe determining how to unequivocally characterise the state of such a system.

This recipe contains information on what may constitute the admissible state of a system, and what cannot constitute such a state. This very information is an important part of the theory. The set of all possible states of a given system is called its phase space. In other words, every point of a phase space is a certain state of the system under consideration. Studying the mathematical structure of the phase space is sometimes a difficult yet very important element of constructing a physical theory, since this structure defines what is possible in a given theory and what exceeds its remit. This is the so-called kinematic part of the theory. As yet, there is no dynamics in it, only, in a way, the setting of the scene on which the dynamics will play out.

The dynamics are supplied by a differential equation defined on a given phase space. Its solutions determine various curves on this space (sometimes, for the sake of simplicity, we say that these curves constitute the solutions of a given equation). On every such curve, the equation also defines its direction. Let us note that such an oriented curve determines the sequence of states through which the system passes, thus determining the evolution of the system compatible with the direction of the curve. This direction can be identified with the direction of the passing of time.

In this context, we should note Hume's well-known and often repeated reservation that in physics we cannot maintain that A causes B, but only that B succeeds A.[172] If A and B are taken to represent the states of a certain system, Hume's view in the domain of the theory of dynamical systems is blatantly false. A dynamic equation not only describes a succession of states, but also 'imposes' on it an authentic dynamics, since it comprises a mechanism of sorts that triggers a true dynamic relationship among the successive states. We say in brief that a dynamic equation not only describes something but also causes what it describes. State B not only succeeds state A, but also dynamically results from it.

A dynamic equation usually has several solutions, i.e. it determines numerous evolutionary routes (curves in a phase space). Which one will be taken by the evolution of the system under investigation? The equation itself does not determine it. From a mathematical point of view, a specific curve in the phase space must be indicated by choosing its initial conditions. In other words, it is necessary to 'tell' the equation from which point in the phase space it should begin counting the evolution. From the point of view of physics, such an initial state should be chosen as the starting point for evolution. A simple example: I am throwing a stone (that can be idealised as a single mass point). A dynamic equation (expressing Newton's law of

---

[172] Hume's argument came down to the claim that experientially we are only made aware of sequences of phenomena, not of their causal mechanisms.

motion) defines a number of curves (in the phase space) along which the motion may take place, but in order to choose a specific curve, I must throw the stone from a certain place and with a certain initial velocity. In this way, I am choosing the initial conditions for a suitable solution.

The entire description above is modelled on classical dynamical systems. Such systems are deterministic, i.e. the choice of a single state unequivocally determines all the states in the past and in the future. But they are not the only dynamical systems, even in classical physics. There are systems in which states are given with a certain probability, for example in the so-called stochastic systems. In this context, the equation does not determine successive states, but may determine the probability distributions of subsequent states. To use a slogan-like expression, although the states themselves are not determined, the probabilities of these states becoming realised are determined. Then, we are dealing with a deterministic evolution of probabilities. Such a situation occurs in the evolution of quantum systems described by the Schrödinger equation. If the quantum system under consideration is for example an electron, its subsequent states can be calculated using the equation only with a certain probability, but the probabilities of successive states (the distribution of probabilities of a given state) are defined by the equation in a deterministic manner.

I would like to draw the Reader's attention to the enormous wealth of dynamical systems. Above, I have introduced their simplest cases and simplest properties, but it can be said without exaggeration that everything subject to change in nature constitutes a certain dynamic system. In particular, every living organism represents such a system.[173]

---

[173] I would like to draw the Reader's attention to the fact that here and in a number of other places I have resorted to a kind of simplification. What I call a dynamical system is both its mathematical model and the associated processes that occur in the physical world that such a model represents. A precise discrimination of these concepts in each and every instance in the text would be tiresome for the Reader, with the danger of misunderstandings being minimal.

5. The dynamics of life

If everything that changes in nature constitutes a certain dynamical system, living organisms are dynamical systems. Moreover, there are strong reasons to suppose that biological evolution as such (on our entire planet) is also a dynamic system, since it can be easily checked that it meets all the essential criteria of a dynamical system. Yet, for the sake of simplicity, let us consider a single evolutionary strand that consists of successive generations descended from a certain individual.

Accordingly, we have a huge phase space – a set of all possible states in which our evolving dynamical system can find itself. This phase space is determined first of all by the laws of physics. The states that would contradict these laws are excluded from the phase space from the very beginning. But physics imposes on the phase space a host of other constraints, since there are numerous situations which – even though they conform to the laws of physics – might exclude the occurrence of a dynamic process. An admissible state must meet certain chemical, biochemical and biological requirements, but even if we consider what is left, we are facing an unimaginably large area of possibilities.

In this huge space of states, the system covers a certain route, goes through successive states, that is, it determines a curve in the phase space. Naturally, it would be naive to claim that we can write an equation (or a system of equations) whose solution would be given by the curve in question. Yet the thing is that such curves, or evolutionary paths, are, in fact, determined by the laws of physics, and, in principle, every law of physics is expressed by a differential equation (or a system of equations of this kind). Thus, somewhere 'in the background' of the process of evolution, dynamic equations are concealed.

However, without a shadow of a doubt, the dynamical systems responsible for biological evolution must possess certain features that make them different from many other systems of this kind. First of all, they must be creative and for that reason, they must be dynamical non-linear systems in states far from equilibrium. Only such systems can be really creative.

The fact that they must be dynamical systems in states distant from equilibrium is rather obvious, since an equilibrium means death, the end of all creative processes. But a creative system must also be a non-linear one, i.e. the equation responsible for its dynamics must be a non-linear differential equation (or a system of equations of this kind). Non-linear differential equations differ from linear ones in that in the latter, the sum of two solutions is a new solution, whereas in the non-linear ones it is not the case. The situation in linear equations can be compared to a building made of blocks. It can be very intricate, but is never anything more than a sum of its parts. Non-linear equations admit of structures that cannot be decomposed into the sum of their constituent parts. This also applies to the temporal direction: the present state need not be a sum of the parts of the previous states even though they may be otherwise rearranged. For this very reason, creative processes are possible in which a truly new structure comes into being – one that has not been there before.

To summarise this part of our considerations, we should once again say that biological evolution has all the hallmarks of a non-linear dynamic system found in states distant from equilibrium.

6. Natural selection

Until now, we have not mentioned one important mechanism of evolution, namely natural selection. Mutations occur at random and are usually disadvantageous, but in the process of interaction with the environment they tend to become eliminated, unlike the advantageous ones, which are favoured. As a result, they accumulate in subsequent generations and combine to form the driving force of further advantageous changes. This mechanism is certainly considerably more complicated and has various variants, but – generally speaking – that is what it consists in. Do such complicated processes somehow compare with the theory of dynamical systems? Not only they do, but outside the structure of a dynamic system, they would not be able to operate at all.

Creative systems must be open, they must exchange energy with their environment (by absorbing low-entropy energy and expelling high-entropy one). But openness exposes such systems to various stimuli originating from the environment. If these stimuli are strong enough, they may destroy the evolution of the system or even the system itself. A sudden stimulus may also only slightly modify the evolution of a system. Then nothing especially interesting happens. Yet the impact of a stimulus does not depend exclusively on its force, but also on how the system reacts to the stimulus. The most interesting things happen when small stimuli cause considerable changes in its evolution.

Sensitivity to stimuli may occur even at the level of the initial conditions of a dynamic equation. We call it the sensitivity of the equation to the fluctuations (perturbations) of the initial conditions. If the sensitivity permits small fluctuations of initial conditions to cause considerable changes in subsequent evolution, such a system is called a system with dynamic chaos. Then, even if the dynamic equation is completely deterministic (just as is the case in classical mechanics), the system remains completely unpredictable. If we knew its initial conditions with an infinite accuracy, we could unequivocally predict the future evolution of the system, but such a knowledge of initial conditions is, even in principle, impossible. In systems with a dynamic chaos, even the least inaccuracy in our awareness of its initial conditions obviates the possibility of unequivocal predictions. Details of this mechanism depend on the mathematical structure of the dynamic equations. The abundance of possibilities is huge. Because a dynamic chaos underlies the processes leading to the emergence and evolution of progressively more complex structures in nature, the abundance of complex structures around us testifies to the wealth of possibilities contained in dynamic equations which are responsible for these processes.[174]

To be sure, the initial conditions are very important for the future evolution of a system, after all, a lot depends on the point of departure. But the subsequent interactions with the environment are also of key importance to the entire process. As we know, an open system is exposed to outside stimuli. There are also fluctuations of the environment. Here also a lot depends on the structure of dynamic equations. Sometimes the system remains stable because the external fluctuations are small, i.e. under the influence of such fluctuations, the system imperceptibly deviates from its evolutionary path, soon to return to it. Yet sometimes in the solutions of dynamic equations (i.e. on curves in the phase space that represent the solutions) there are points of instability. In any of these, the system is especially sensitive to fluctuations coming from the environment. A small fluctuation, which in a period of stability would be completely innocent, may knock the entire system off its previous evolutionary path and direct it towards a completely different one (a solution of the dynamic system or an admissible curve in the phase space). Usually, there are numerous admissible paths (solutions of a dynamic equation) and where the system eventually proceeds, depends on the kind of the external fluctuation. In the theory of dynamical systems, such a point of instability is called a bifurcation point: a system that finds itself at an intersection has numerous routes to choose from, yet the choice is not decided by the system itself, but by the external fluctuation that will push it in one direction or another. Along the new evolutionary path, there are states which would have been unattainable on the old path. This is how 'qualitative leaps' in the evolution of the system occur.

---

[174] A more detailed and in-depth treatment of the theory of the formation of complex structures in the Universe can be found in P. Davies, *The Cosmic Blueprint*, Templeton Foundation Press, Philadelphia – London 2004.

The mathematical abundance of different possibilities, even in relatively uncomplicated dynamical systems, is enormous. Despite the fact that so far we have managed to study only a small subset of all dynamical systems existing in mathematics, we are astounded by their variety and creative possibilities.

After this somewhat longish lecture, we will have no serious difficulties in accepting the assumption that natural selection, a very significant factor in biological evolution, becomes part and parcel of the strategy of dynamical systems by way of their sensitivity interacting with the fluctuations of initial conditions and the impacts of the fluctuations originating from the environment. The point is, of course, to formulate a principle, and not to make sure that we can actually mathematically model the processes associated with biological evolution, even though various handbooks of mathematical theory of dynamical systems contain chapters devoted to differential equations that model such processes.[175] I do not want to say that all biology can be reduced to the theory of dynamical systems, but that biology does not operate above and beyond the laws of physics, on the contrary, it is based on them and applies them in practice. The laws of physics provide the foundations and key strategies. Biology fills them with its own contents and enriches them with its own specificity. Natural selection, as expounded by the contemporary theory of evolution, is something specifically biological, but it would not be able to work outside the strategy of dynamical systems. Biology does not work against physics, but is derived from it.

---

[175] Cf. e.g. Chapter 12 of a well-known coursebook: M.W. Hirsch, S. Smale, *Differential Equations, Dynamical Systems, and Linear Algebra*, Academic Press, New York – San Francisco – London 1974.

Chapter 15

Intelligent Design or the Mind of God?

1. Two strategies

At first sight, both expressions used in the title of this chapter mean exactly the same, after all, how can the Mind of God be anything different from Intelligent Design? – nevertheless, they are divided by a huge gap in meaning. This gap is an artificial creation, an outcome of purely historical developments, yet today it may no longer be disregarded. In previous chapters, we have devoted a lot of attention to Intelligent Design, its origin and most important claims. As we remember, one of its main strategies relies on the treatment of chance as a rival to Intelligent Design. If we encounter an event whose occurrence is improbable, we may attribute it either to 'pure chance' or to the work of the Intelligent Designer, but if the probability of such an event is 'indeed extremely low,' attributing it to pure chance would seem irrational, so it can only be treated as a trace left by the Designer.

The strategy of the Mind of God is totally different. The very expression was coined by Einstein. He used to say that the only thing he wished to know was the Mind of God. It should be understood in the sense that by creating the world, God realised a certain Creative Intention. Since science endeavours to understand the world, it does nothing else but tries to decipher this Intention. For now, it is not important whether God is understood as the Christian almighty 'Creator of heaven and earth,' or more like Einstein, who was inclined to identify the Mind with God. At any rate, the Mind of God embraces everything in the Universe. If indeed there are random events in the Universe, even extremely improbable ones, they also constitute part of the Mind of God, because everything that constitutes a challenge to science is part of it.

In this chapter, by way of summing up our earlier discussion and giving it its due corollary, we will try to gain a better insight into 'the philosophy of the Mind of God' and to understand the part played in it by random events.

2. Necessity and chance

The notion of chance has already been discussed at sufficient length in previous chapters. The concept of necessity also has a number of meanings and has been variously understood in the history of philosophy. In our present considerations, we will need its more restricted understanding. We agree that something is necessary if it results from the operation of a certain law of nature (for the most part, we will be talking about the laws of physics). Naturally, a question immediately arises as to whether or not the laws of nature themselves are necessary. Or could they be different from what they are? But let us leave aside this question and other similar ones. Besides, the issue of necessity interests us at present only marginally as a correlate of the notion of chance. The best solution would be to go straight to an issue where both concepts cooperate, because this is our first important observation: chance and necessity not so much oppose as cooperate with each other.

As we already said in the previous chapter, every living organism is a non-linear, open and dynamical system that finds itself in states distant from equilibrium. In every dynamic system of the kind (not only in living organisms), two elements can be distinguished: the element of necessity – i.e. the operation of a certain law of nature, and the incidental, random element – different fluctuations that assault the system both at the level of initial conditions and later, during its further evolution. Laws of nature are primarily expressed in terms of mathematical equations, but also as entire mathematical structures associated with them, i.e. the phase space with all its geometric architecture, all the possible solutions and the way they arrange themselves in the phase space (phase space topology). All of this works with mathematical precision, giving mathematicians who investigate dynamical systems both ample aesthetic experience and numerous challenges that they must face up to. On the other hand, the random element is constituted by different fluctuations of the environment to which the system is exposed and to which it sometimes succumbs. They are not part of any law of nature represented by the dynamical system, but originate from outside. For that very reason, they are random. From within a given dynamical system, it is impossible to predict when and which fluctuation will strike. This, however, does not mean at all that fluctuations derive from a 'twist of fate', that they appear without any justification, that they constitute a 'breach in the rationality of the world.' They also result from the operation of different laws of nature.

Embryogenesis is doubtless a dynamical system or, more precisely, a dynamical subsystem of a larger dynamical system which is its mother organism. A quantum of cosmic radiation hits one of the embryo cells causing a mutation. From the perspective of the embryo, this is unquestionably a random event. Before the quantum hit the cell, the dynamic system of the embryo contained no information that something like that would happen. But then, cosmic radiation was born somewhere in the depths of the Universe on the strength of the laws of physics and it was those laws of physics that made the quantum travel the way it did, and, after some time, hit a cell that was in its path (the fact that these laws were probabilistic [quantum], does not alter the situation significantly). We may, once again, invoke the example quoted by St Thomas Aquinas after Aristotle (cf. Chapter 12.4) about the man who went to the market for a reason and met – by chance – his creditor who also went there, but for another reason. Using St Thomas' language, it was an intersection of two causal chains in which there was nothing accidental (since both men went to the market with their specific respective objectives), and chance was born only at the intersection of those chains.

Fluctuations that attack a dynamic system result from the operation of different laws of physics and in many situations they need not be quantum in character, and thus probabilistic, laws. Their randomness is determined by the fact that they are external with respect to a given dynamical system that they cannot be predicted from within such a system. And this is our second important observation: an event is random only with respect to a given law of physics. With respect to a different law of physics, it may even be a strictly deterministic consequence of its operation.

Random events thus understood, although they are something external with respect to a non-linear system under consideration, play a significant role in its structure, since 'cooperation' with random events constitutes an inherent part of the structure of such a dynamic system. Let us remember (cf. the previous chapter) that at a certain stage of its evolution, a given system may be entirely insensitive to the impacts of (not too large) fluctuations of the environment, whereas at bifurcation points it suddenly becomes very sensitive to them. And this fact is no longer something external to the system. For a multitude of dynamical systems, the structure of their phase spaces can be calculated and plotted by a computer (by means of the appropriate software). Such plots are called phase portraits. Such a portrait offers information on which parts of the solutions (or evolutions) are stable with respect to the operation of fluctuation and where the bifurcation points are to be found. In simpler cases, it may be an instructive exercise for students.

The choice of the initial conditions for a dynamical equation is also a random element with respect to a given system. When we solve dynamical equations, the choice of initial conditions is, in principle, facultative (for a specific kind of dynamical systems the initial conditions must meet the so-called 'constraint equations'). Obviously, unless it is only an exercise in mathematics, the initial conditions must be compatible with the physical situation under consideration, but this does not alter the fact that they are random with respect to the dynamical system itself. If I throw a stone, it will surely fly along the trajectory that constitutes the solution to the Newtonian motion equation, but the choice of the starting point of the throw and its initial velocity depend on my decision. In processes that occur in nature, the initial conditions are 'imposed' by other physical processes, which are random with respect to a given system. For example, a stone avalanche in the mountains descends in accordance with the statistical laws of classical mechanics, but it can be started by some other process – let us say, by a tree felled by the wind. But – and this is important – without the imposition of initial conditions, a dynamic system would not be able to operate, the equation 'would not know' which solution to choose, and thus the process modelled by a given dynamic system would not start. Accordingly, in the structure of a dynamical system there are certain 'places' left for random events to operate. These places constitute an important part of the structure of a dynamical system.

Let us take the liberty of visualising our considerations thus far. Let us imagine all the laws of physics operating in the Universe as a great network or grid. The image is justified inasmuch as the laws of physics are interconnected in various ways, which results in a great and intricately built structure. But this network contains certain 'empty spaces' left for random events to operate. Without them, the entire structure would not be able to function. Moreover, there are as many 'empty spaces' as are needed – neither more nor fewer – for the structure to work effectively.

Not surprisingly, every comparison disappoints in one respect or another. As we have seen, random events do not operate in 'empty spaces,' but rather at the 'intersections' of different laws of nature, but maybe the image of a network with such 'empty spaces' appeals better to our imagination.

Let us formulate the conclusion from these considerations as our third important observation: random events are not 'foreign bodies' in the structure of the laws of nature, but constitute their important elements.

This observation must inevitably be supplemented by the next: the presence of 'empty spaces' in the structure of laws of physics for the operation of random events is a necessary condition for the processes leading to the growth of complexity to occur, in other words, to make biological evolution in the Universe possible. As we have seen, the openness of non-linear dynamical systems to the operation of random fluctuations is essential if these systems are to be creative, i.e. if they are to produce significant novelties in the course of their evolution. Likewise, the contemporary complexification theory states that the necessary condition for complexification to occur is a certain touch of dynamic chaos in dynamical systems. Dynamical chaos presupposes sensitivity to small fluctuations of initial conditions, which is – as we have seen – a random element with respect to a given dynamic system.

In summary, let us offer yet another remark: all the observations presented above are strictly mathematical in nature in the sense that in all the simpler cases (i.e. those that we can mathematicise) they constitute conclusions that stem from mathematical theories and models. We are entitled to believe that even those cases that exceed the capacity of our present-day mathematical tools, are also mathematical. The entire history of physics is full of examples of phenomena which were considered impossible to mathematicise, and yet after some time, they were fully mathematicised.

3. Should biology arouse metaphysical emotions?

Philosophising biologists and philosophers with an interest in biology who have positivistic inclinations, often express the opinion that 'there is nothing metaphysical in biology,' and that all can be understood and explained without recourse to any 'external factors.' When nailed down, they say that this 'understanding and explanation' consists in a reduction of biology to physics (if chemistry is to be understood as part of physics). Bearing in mind the discussion in previous chapters, we are inclined to agree with these views. True, but these views need a very important caveat. For what does it mean 'to reduce'? There are many different reductionisms.[176] The most popular understanding of reductionism is described by Paul Davies as follows:

> The behaviour of a macroscopic body can be reduced to the motion of its constituent atoms moving according to Newton's mechanistic laws. The procedure of breaking down physical systems into their components and looking for an explanation of their behaviour at the lowest level is called *reductionism*, and it has exercised a powerful influence over scientific thinking.[177]

Here, Davies explains the meaning of reductionism using atoms and Newtonian mechanics as an example (the so-called mechanistic reductionism). Today atoms should be replaced with subatomic particles or other even their more fundamental components, and Newtonian mechanics with quantum mechanics, quantum field theory or another even more basic physical theory (yet to be developed), but the core idea remains the same: one should break down the whole into its more basic components and to explain the behaviour of the whole via the behaviour of its constituent elements.

Reductionism thus understood has rendered invaluable service to the advancement of science, but today we know that it does not explain everything (in principle, it fully operates only in linear dynamical systems). Certainly, it does not fully explain the passage from the inanimate world to the animate one and the ensuing evolution of

---

[176] Cf. e.g. no. 1 volume II (1997) of the journal "Foundations of Science"; the entire issue is devoted to the problem of reductionism.

[177] P. Davies, *op. cit.*, pp. 12–13.

life. As we said in the previous chapter, evolutionary processes are non-linear ones with a strong sensitivity to small perturbations of the initial conditions (deterministic chaos), and afterwards to the influence of external fluctuations. We also remember that the structures of such systems cannot be reduced to the sum of their parts (because they are non-linear), moreover, their future states may contain important new developments which did not exist in previous states, so the successive states cannot be deduced from the previous ones. The emergence of such novelties does not require any external intervention apart from the laws of physics, but those laws must be such that permit the launching of non-linear processes far removed from states of equilibrium, and must possess properties (discussed in previous chapters) indispensable to the triggering of increasing complexity.

In the context outlined above, we say that a new quality arose from the previous conditions by way of emergence. Emergence has been extensively discussed in literature.[178] We already know a lot about emergence thanks to the findings by different exact sciences such as non-linear thermodynamics, theories of chaos and complexification, theory of evolution, cognitive science,[179] but the philosophical import of emergence is still an object of an animated discussion.[180] It seems important to note that in emergent processes not only do new qualities arise, but also at a higher level at which these new qualities operate, new regularities begin to apply (some even call them new laws) that emerge from laws governing the lower levels. Sometimes these new regularities are called the organising principles.[181] They are holistic in scope, that is, they also coordinate even remote elements of a given level.

---

[178] Cf. R. Poczobut, *Między redukcją a emergencją (Between Reduction and Emergence, in Polish)*, Wyd. UWr, Wrocław 2010. The work includes an extensive bibliography.

[179] A very good discussion of these results can be found in the book by Paul Davies, *op. cit.*.

[180] Cf. e.g.: *Struktura i emergencja (Structure and Emergence, in Polish)*, M. Heller, J. Mączka, eds., OBI – Biblos, Tarnów – Kraków 2006.

[181] Cf. Chapter 11 in P. Davies' book.

The possibility of emergence is intricately interwoven into the laws of physics. A small modification of laws or values of constants and of other parameters that characterise the laws would cause the Universe to become either a complete mess or a frozen order similar to crystal structure – beautiful, but always the same. An infinite number (at any rate, in the practical sense) of other, intermediate scenarios is also possible that stipulate that the Universe would be subject to some changes, but would remain forever closed to more complex structures. Let us imagine that the Universe has no property of frequency stability (cf. Chapter 7.7). In such a case, probability calculus and statistics would not refer to the world and all the mechanisms based on them, so important for the organisation of the world, together with natural selection, would not work. But the laws of physics permit the existence of non-linear dynamical systems characterised by appropriate sensitivity to initial conditions and external fluctuations – and this is also an admirable property of the laws of physics, which they 'very easily' might not have.

There exists extensive literature devoted to the so-called anthropic principles. It collates various properties of the laws of physics, values of physical constants, the initial conditions and various parameters characterising the Universe etc. whose even slightest modification would cause the Universe to be unfit for life. At first, anthropic principles were objects of interest as it were for their own sake,[182] and subsequently the issues of the anthropic principles was considered almost exclusively in the context of the concept of the multiverse (cf. Chapter 13.3). We will not embark on a review of these issues,[183] it is enough to realise that everything ultimately depends on the laws of physics and, possibly, on the initial conditions of the Universe.[184]

---

[182] A classical work from that period is J.D. Barrow, F.J. Tipler, *The Anthropic Cosmological Principle*, Clarendon Press, Oxford 1986.

[183] I devoted to this issue Part Two of my book: *Ultimate Explanations of the Universe*, *op cit.*

[184] I have written 'possibly,' because as yet we do not know how these came about or even whether they were necessary at all. We hope to find out about it from the final theory of physics once it has been discovered at long last.

And this is where metaphysical emotions appear. Life emerged by way of evolution. Biological evolution is a strand in the evolution of the Cosmos. And the evolution – both cosmic and biological – is an emergent process... Did it have to turn out this way? And this is what arouses emotions in those who have the capacity to wonder.

4. The Mind of God

Einstein wanted only to understand the Mind of God – the Intention that God built in the work of creation. Einstein had great ambitions, he wanted to discover everything and express the Mind of God in a single formula. Although we try very hard and spare no efforts, we are still very distant from realising his ambitions. Even if today's scientists do not always wish to admit this, they also share in Einstein's aspirations. Doubtless, physics has made great advances and we have the right to say that we have managed to decipher – quite sizeable, it seems – chunks of the Mind of God. And it is not about arranging pieces of a mosaic from until now quite chaotically scattered pebbles, because we now know that this is not a mosaic, but a very intricately composed structure. If here and there small pieces appear to fit together (as in a mosaic), it is only because the structure under consideration – at least in the areas that we know – has a property that enables us to approximate it using simpler structures (just as a small area on the surface of a sphere can be approximated by a plane).

If physicists indeed step by step reconstruct the Mind of God, the entire history of physics proves that God's thinking is mathematical. Naturally, it would be naive to suppose that God uses definitions from our manuals and applies our theorems to determine the relationships amongst the different subrealms of His Intention, it is doubtless a result of His generosity (or absence of malice, as Einstein used to say[185]) that the definitions and theorems formulated by us somehow correspond with His Mathematics. We consider our mathematics to be perhaps the most beautiful product of human rationality. Therefore we are entitled to attribute the feature of rationality – rationality *par excellence* – to the Mind of God. However, it is not a rationality that always conforms with our mental habits, even those that we are inclined to consider obvious. That is why we so laboriously uncover the pieces of God's Mind. It must be admitted that as we do it, we manage to rectify the standards of our rationality, although it does not come to us all that easily. Almost at every turn in our search, we must be ready for unexpected turnabouts which only with the benefit of hindsight reveal to us their own logic.

In this book, I have tried to show the evolution of the concept of chance that it had to undergo before it became integrated into the logic of the Mind of God. Until it entered into the context of the mathematical probability theory, it appeared to be a breach of rationality, at least it was thus understood at the time. This impression remained in force for some time, even seemed to become more fixed, since probability calculus was a response to the needs of games of chance, and we tend to perceive gambling as a nonsensical exposure to losing everything. Only when probability calculus became established and began to interact with other mathematical structures, did it turn out that in respect of its rationality, it was not much different from other mathematical structures. And when it grew to be part of measure theory, it became completely similar to those structures. Applications in physics demonstrated its operation in the Universe. Moreover, not only is it interwoven in the rationality of the Universe, but also constitutes an indispensable element that permits the system of laws of nature to operate at all. To illustrate my point, in the network of laws of physics there are 'empty spaces' for the operation of chance and there are exactly as many of them as are needed for the laws of phys-

---

[185] Einstein's famous saying: "God may be sophisticated, but He is not malicious". Sophisticated – because He uses sophisticated mathematics, not malicious – because He made it partly available to us.

ics not only to be abstract patterns, models or formulae, but also to govern specific processes in nature. Without those 'empty spaces' for the operation of chance, very intricately interwoven into the network of laws of nature, those equations would remain abstract formulae describing only possible behaviours (rather than those that occur in the real world). Obviously, these equations would have their specific solutions, but only the adoption of initial conditions (or boundary conditions) selects one of the available solutions and permits them to model a specific process.

Some 'empty spaces' in the network of the laws of nature for the operation of chance fulfil yet another function, namely they make it possible for truly creative processes to occur in the Universe. What I have in mind is the cooperation of fluctuations coming from the environment with non-linear dynamical processes. From the perspective of a given dynamical system, these fluctuations are random elements, yet they make it possible for creative processes to occur in nature. Thanks to this strategy, natural selection is possible. Without it there would be no biological evolution. Those who would like to eliminate random events from our world, do not realise that without chance our world would cease to function.

5. The game of chance called life

The macroscopic world is full of random events that only lie in wait to take advantage of 'empty spaces' in the network of laws of nature and do their job. This is our macroscopic world, and we who live in it often behave like a bull in a china shop. Every minute or even more often, our actions initiate causal sequences which would never occur otherwise. From the vantage point of the world, they are entirely random. A good question: do they fit in 'the empty spaces' which the laws of nature leave for chance in order to be able to operate themselves? Many of them surely do. When on a trip in the mountains I kicked a stone that started an avalanche, I only imposed the initial conditions on an appropriate equation that afterwards did its job (by means of various other boundary conditions and external stimuli). This example is quite characteristic in that our numerous actions – as opposed to the strategy of small fluctuations in non-linear dynamical systems – trigger causal chains which are not creative, but destructive. Even if all our actions fit in the 'empty spaces' of the network of the laws of nature, most of them cannot be treated as small disturbances in the ordinary course of nature. They constitute very brutal invasions, still nature somehow tolerates them, showing admirable stability in the face of our actions. To be sure, this stability has its own limits – we are capable of destroying our planet, but if this occurred, 'the rest of the Universe' would not even notice. We are not an especially important detail in the structure of the Whole.

However, from the perspective of an individual, matters look quite different. From beginning to end, our lives are entangled in an 'orgy of random events.' We often say that the only non-accidental thing is that if we were born, then we must die. But the circumstances of our births and deaths are purely accidental. Somebody's father may have missed a train and it was how he met his future wife. Then they had a daughter... Somebody else may have been passing under a makeshift roof, was struck by a falling brick and killed on the spot. But we do not have to resort to such important events (in people's lives). All our days are filled with subconscious decisions and gestures that we perform without reflection; they may remain insignificant or trigger consequential chains of events. Our philosophy of life does not invoke the network of laws of nature or mathematical equations. For daily use, we have our own definitions of random events or – more often than not – go by our intuition and an instinct of sorts. Quite often, what we call random events refers to what we do not expect or what surprises us, because we lack appropriate knowledge to be able to predict them, but even then we do not apply the Bayesian evaluation criteria (cf. Chapter 6.8), preferring to make do with our intuition. For the most part, in our lives we behave like the participants in a game of chance, who, even if they heard about the works of Pascal, Fermat and Bernoulli, ignore them in the calculation of their chances.

Does it all fit in with the Mind of God for the Universe, that is, the Great Mathematical Matrix to which the Universe is subordinated? From perspective of natural theology (i.e. if we do not invoke a religious revelation), one may only say that if man has a free will, then the Great Matrix must provide for the possibility of such free will within the Universe. If we have learnt anything from the history of science, we can state that the free will does not violate the mathematical structure of the world (the Great Matrix), but is 'superimposed' on it. The reasoning here is similar to 'retroactive inferencing': if there is life in the Universe (at least on a single planet), then the laws of physics must provide for processes leading to biogenesis. The difference comes down to the fact that in the case of life, we already know the physical basis for these processes, whereas in the case of consciousness, intelligence and the free will (these three elements appear to be closely related) we have only a series of more or less reasonable suspicions.

6. The Great Matrix

Above, I have used the metaphor of the Great Matrix of the Universe. Partly it is due to the fact that I did not wish to abuse Einstein's expression of the Mind of God, but essentially, both metaphors refer to the same thing. Irrespective of whether we interpret it in the spirit of religious faith in God or in the sense of a set (but is it really a set in the technical sense of the term?) of regularities that govern everything that exists, we must agree that this Matrix is really Great. We would probably be entitled to say Infinite. And here again we return to the metaphysical amazement: we have managed to decipher this much! If this is to be an authentic metaphysical amazement, the statement should not so much recognise our intellectual abilities as express our joy at the fact that the Great Matrix has revealed so many of its secrets to us, combined with an awareness that the deepest ones can only be sensed. We may ask about them, though we are not certain if we are using the right words.

In this metaphysical amazement there is also something of an aesthetic experience. The Great Matrix – in its part that we already know – is not only Beautiful, but also appears to be the Prototype, the Blueprint for all that we consider to be beautiful.

Yet we should not forget about our capacity to get to know it, because, in the end, our mind also constitutes part of the Great Matrix (the Mind of God), moreover, a part in which the Great Matrix possesses an awareness of itself. Our mind, communing with small portions of the Matrix available to it participates in a Great Rationality – a Rationality that not only embraces it but exceeds it. In this book, I have tried to describe the struggles and attempts of numerous generations of thinkers to penetrate certain areas of the Great Matrix – areas of special difficulty, because they seemed to indicate that the concept of the Great Matrix collapses in confrontation with the destructive force of chance. We already know the result of these struggles. Random events do not constitute a defeat of the Matrix, but represent its very subtle strategy. The power of the Great Matrix consists in the fact that anyone who wants to show that it does not exist or to constrain its effectiveness – as long as they are using rational arguments – must make use of the resources supplied by the Matrix itself.

...as I was writing these words, my glasses fell off my face. Let this random event serve as an end to this book.

Tarnów, December 1, 2010

Subject index

Names index